Other Titles in This Series

(*Continued in the back of this publication*)

Canard Cycles
and Center Manifolds

MEMOIRS
of the
American Mathematical Society

Number 577

Canard Cycles
and Center Manifolds

Freddy Dumortier
Robert Roussarie

May 1996 • Volume 121 • Number 577 (first of 4 numbers) • ISSN 0065-9266

American Mathematical Society
Providence, Rhode Island

1991 *Mathematics Subject Classification.*
Primary 34E15, 58F14; Secondary 58F21, 58F30, 58F36, 34C30, 34E20, 34C20.

Library of Congress Cataloging-in-Publication Data

Dumortier, Freddy.
  Canard cycles and center manifolds / Freddy Dumortier, Robert Roussarie.
    p.   cm. – (Memoirs of the American Mathematical Society, ISSN 0065-9266; no. 577)
  "Volume 121, number 577 (first of 4 numbers)."
  Includes bibliographical references.
  ISBN 0-8218-0443-X (alk. paper)
  1. Boundary value problems—Asymptotic theory.   2. Perturbation (Mathematics)   3. Bifur-
cation theory.   I. Roussarie, Robert H.   II. Title.   III. Series.
QA3.A57   no. 577
[QA379]
510 s–dc20
[515'.352]   96-2234
                                                                                                    CIP

Memoirs of the American Mathematical Society

This journal is devoted entirely to research in pure and applied mathematics.

Subscription information. The 1996 subscription begins with Number 568 and consists of six mailings, each containing one or more numbers. Subscription prices for 1996 are $391 list, $313 institutional member. A late charge of 10% of the subscription price will be imposed on orders received from nonmembers after January 1 of the subscription year. Subscribers outside the United States and India must pay a postage surcharge of $25; subscribers in India must pay a postage surcharge of $43. Expedited delivery to destinations in North America $30; elsewhere $92. Each number may be ordered separately; *please specify number* when ordering an individual number. For prices and titles of recently released numbers, see the New Publications sections of the *Notices of the American Mathematical Society*.

Back number information. For back issues see the *AMS Catalog of Publications*.

Subscriptions and orders should be addressed to the American Mathematical Society, P. O. Box 5904, Boston, MA 02206-5904. *All orders must be accompanied by payment.* Other correspondence should be addressed to Box 6248, Providence, RI 02940-6248.

Memoirs of the American Mathematical Society is published bimonthly (each volume consisting usually of more than one number) by the American Mathematical Society at 201 Charles Street, Providence, RI 02904-2213. Second-class postage paid at Providence, Rhode Island. Postmaster: Send address changes to Memoirs, American Mathematical Society, P. O. Box 6248, Providence, RI 02940-6248.

TABLE OF CONTENTS

ABSTRACT

In this paper the "canard phenomenon" occuring in Van der Pol's equation $\varepsilon\ddot{x} + (x^2 + x)\dot{x} + x - a = 0$ is studied. What happens is that for sufficiently small $\varepsilon > 0$ and for decreasing a, the limit cycle created in a Hopf bifurcation at $a = 0$ stays of "small size" for a while before it very rapidly changes to "big size", representing the typical relaxation oscillation. A geometric explanation and proof of this phenomenon is given using foliations by center manifolds and blow-up of unfoldings as essential techniques. The method is general enough to be useful in the study of other singular perturbation problems.

Key words and phrases : *singular perturbation problem, geometric explanation, canard phenomenon, Van der Pol's equation, Hopf bifurcation, limit cycle, limit periodic set, normal form, center manifold, foliation, blow-up of unfolding, desingularization, foliated local vector field, family rescaling, phase-directional rescaling, Abelian integrals.*

ix

INTRODUCTION

The "canard (duck) phenomenon", occuring in Van der Pol's equation

$$\varepsilon\ddot{x} + (x^2 + x)\dot{x} + x - a = 0 \qquad (1)$$

was first described and studied in [BCD] by means of non-standard analysis. Essentially it says that for sufficiently small $\varepsilon > 0$ and for decreasing a, the limit cycle created at $a = 0$ stays for a while of "small size" before it rapidly changes to "full size" (see below for an accurate definition). This change together with the occurrence of the intermediate shapes (called "canards" in [BCD]) essentially happens in a small interval $[a_1(\varepsilon), a_2(\varepsilon)]$ of length $|a_2(\varepsilon) - a_1(\varepsilon)| = O(e^{-K/\varepsilon})$, for some $K > 0$, when $\varepsilon \downarrow 0$. We will give a more precise statement below.
A standard treatment, based on asymptotic analysis, was done in [E].

In this paper we want to explain and prove the canard phenomenon in a (standard) geometric way. The main ingredients of our method are center manifolds and desingularization (blowing up) of families of vector fields. As such the method is general enough to be used in the study of other analogous singular perturbation problems.
The desingularization will transform the 2-parameter family of 2-dimensional vector fields (1) into a 4-dimensional (local) vector field, having only elementary singularities, and which can be studied as a 1-parameter family of 3-dimensional vector fields. This technique was introduced in [R] and [DeR] and should be seen as a generalization of the usual rescaling method. We refer to [DR] for a nice application of the method to the study of a local problem and to [D] for a treatment of the Andronov-Hopf bifurcation, merely done for didactical purposes.
After desingularization the limit cycle(s), for ε near zero, will be obtained by transverse intersection of center manifolds associated with the (desingularized) lines of normally hyperbolic singularities.

In chapter 1 we first transform the initial equation (1) to a family of planar vector fields, depending on the parameters $(\varepsilon, a) \in I\!\!R^+ \times I\!\!R$, and having a curve of zeroes for $\varepsilon = 0$. We then give a precise formulation of the canard phenomenon in two different ways.

Received by the editor March 8, 1994.

1

In the second chapter we use the global desingularization to produce a new system $\bar{X}$, having only elementary singularities.

In chapter 3 we introduce C^k-normal forms of $\overline{X}$ near the non-isolated normally-hyperbolic singularities in order to establish the existence of "foliations" of center manifolds. We also prove some transversality properties for the leaves of such foliations.

In the final chapter we will deduce the canard phenomenon from the intersection properties previously established.

In the appendix we give a short proof - due to Chengzhi Li - of a statement used in theorem 18.

Let us also mention that after finishing this paper a short survey of the proof, including a presentation of the essential geometric techniques, appeared in [D].

1 Statement of the result : the "canard phenomenon" for the singular Van der Pol equation

Let us recall that we want to study the second order scalar differential equation as given in (1) :

$$\varepsilon\ddot{x} + (x^2 + x)\dot{x} + x - a = 0$$

for $a \in I\!R$ and $\varepsilon > 0$ but small.

We first reduce it to a first order equation in the (x, Y)-plane where $Y = \varepsilon\dot{x}$.
We have :

$$\begin{cases} \varepsilon\dot{x} = Y \\ \dot{Y} = a - x - \dfrac{Y}{\varepsilon}(x + x^2) \end{cases} \tag{2}$$

Note that, if $\varepsilon \to 0$, a domain $|Y| \leq K$ becomes unbounded in the phase space $(x, \dot{x})$. So it will suffice to study such a domain in the (x, Y)-plane.

Next, we use the Liénard transformation :

$$\begin{cases} y = Y + F(x) \\ x = x \end{cases} \quad \text{with} \quad F(x) = \int_0^x (\xi + \xi^2)d\xi = \frac{x^2}{2} + \frac{x^3}{3} \tag{3}$$

This transformation is a diffeomorphism from the (x, Y)-plane to the (x, y)-plane. It brings the equation (2) to :

$$P_{\varepsilon,a} : \begin{cases} \dot{x} = \dfrac{1}{\varepsilon}(y - F(x)) \\ \dot{y} = a - x \end{cases} \tag{4}$$

Van der Pol's equation, in the form (4), was studied in [BCD] by methods from non-standard analysis. In this context, the cubic curve $L = \{y = F(x)\}$ is the slow manifold and the field given by (4), for $\varepsilon = 0$, is infinitely large and horizontal outside L. All the features of (4) are then described in a precise way, in the context of non-standard analysis, as a "slow-fast" system.

3

Our approach will be different and will remain in the context of "regular" families. We replace the field $P_{\varepsilon,a}$ by $X_{\varepsilon,a} = \varepsilon\, P_{\varepsilon,a}$ (or equivalently, we rescale the time by a factor $1/\varepsilon$) :

$$X_{\varepsilon,a}: \begin{cases} \dot{x} = y - \dfrac{x^2}{2} - \dfrac{x^3}{3} \\ \dot{y} = \varepsilon(a - x) \end{cases} \tag{5}$$

For $\varepsilon = 0$, the vector field $X_{0,a} = X_0 = (y - F(x))\dfrac{\partial}{\partial x}$ has the curve $L = \{y = F(x)\}$ as a curve of zeroes. Outside L, X_0 is horizontal. Observe that, except for the extremal points $n = (-1, \frac{1}{6})$ and $s = (0,0)$, all the points on L are normally hyperbolic singular points. The phase-portrait of X_0 is given in figure 1. We denote by L_i, for $i = 1, 2, 3$, the three open arcs on L, delimited by the points n, s :

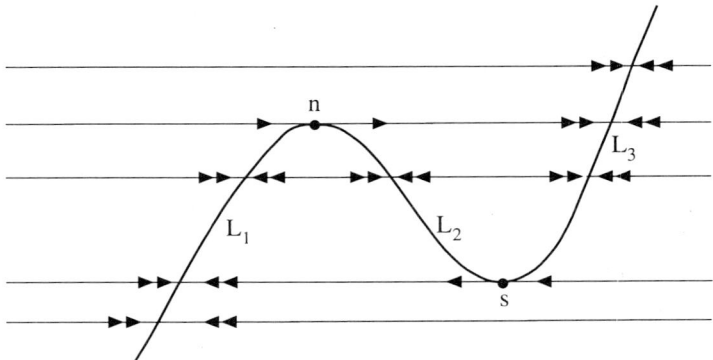

Figure 1
The vector field X_0

We can observe that the equation (5), for each ε, is invariant under the symmetry around the point $x = -\frac{1}{2}$, $y = F(-\frac{1}{2}) = \frac{1}{12}$, $a = -\frac{1}{2}$, given by $(x, y, a) \mapsto (-x - 1, -y + 1/6, -a - 1)$.

So it suffices to consider the value $a \geq -\frac{1}{2}$ and to complete the diagram by the symmetry which exchanges the points n and s.

It is very easy to establish the bifurcation diagram for $X_{\varepsilon,a}$. We suppose that the parameter is in the domain $\varepsilon > 0$, $a \geq -\frac{1}{2}$.

$X_{\varepsilon,a}$ has a unique singular point $p_a = (a, F(a))$. For any $\varepsilon > 0$ the linear part of $X_{\varepsilon,a}$ at p_a is the 2×2 matrix :

$$
DX_{\varepsilon,a}(p_a) = A = \begin{pmatrix} -a - a^2 & 1 \\ -\varepsilon & 0 \end{pmatrix}
$$

Its follows that p_a is an attracting node or focus for $a > 0$ and a repelling one for $a < 0$. In fact p_a is a focus inside the region given by $\varepsilon > \dfrac{1}{4}a^2(1+a)^2$ and this focus changes stability for $a = 0$. Looking at the normal form of the equation along this bifurcation line $H = \{a = 0, \varepsilon > 0\}$ it is easy to see that this bifurcation is a generic subcritical Hopf bifurcation. This means that for a crossing zero and decreasing, there appears a small attracting limit cycle around the repelling focus.

To complete the bifurcation diagram we use the following fact concerning the possible limit cycles in $X_{\varepsilon,a}$:

$X_{\varepsilon,a}$ has at most one limit cycle and when it exists it is hyperbolic and attracting

This fact is common knowledge concerning Van der Pol's equation. It can be found in [LMP] but it may also be deduced from a general result of W. Coppel concerning Liénard equations [C].

It implies that for $a > 0$ there exist no limit cycle and the basin of attraction of p_a is the whole plane. For $a < 0$ there exists just one limit cycle $\Gamma_{\varepsilon,a}$, which is attracting, having as basin of attraction $I\!\!R^2\backslash\{p_a\}$. The study in the whole parameter space is completed by the above mentioned symmetry adding a second line $H' = \{a = -1\}$ of Hopf bifurcations. The bifurcation diagram is shown in figure 2. The dotted line is not a bifurcation line, but represents the transition from focus to node.

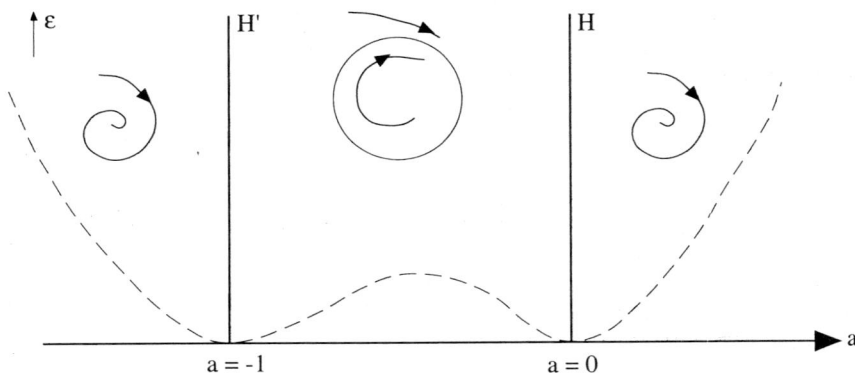

Figure 2

Bifurcation diagram for $X_{\varepsilon,a}$

Now an interesting question is :

What happens to the limit cycle $\Gamma_{\varepsilon,a}$ when $\varepsilon \to 0$?

A possible limit (for the Hausdorff distance between the compact sets in $I\!R^2$) of a sequence of limit cycles is called a *limit periodic set* (see eg. [R])

So the question above deals with the possible limit periodic sets existing for X_0. It is not hard to see that they are all shown in figure 3.

The exact content of the question deals with the way that limit periodic sets will be approached by 1-parameter families of limit cycles $\Gamma_{\varepsilon,\gamma(\varepsilon)}$ when considering 1-parameter families $X_{\varepsilon,\gamma(\varepsilon)}$.

As we keep $a \geq -1/2$ we will only have to deal with the cases (1) - (5). The other ones are their symmetric counterparts.

Let us now introduce some terminology.

A limit periodic set like in (1) is called "small" and is denoted as Γ_0, while one like in (5) is called "big" and denoted as Γ_B.

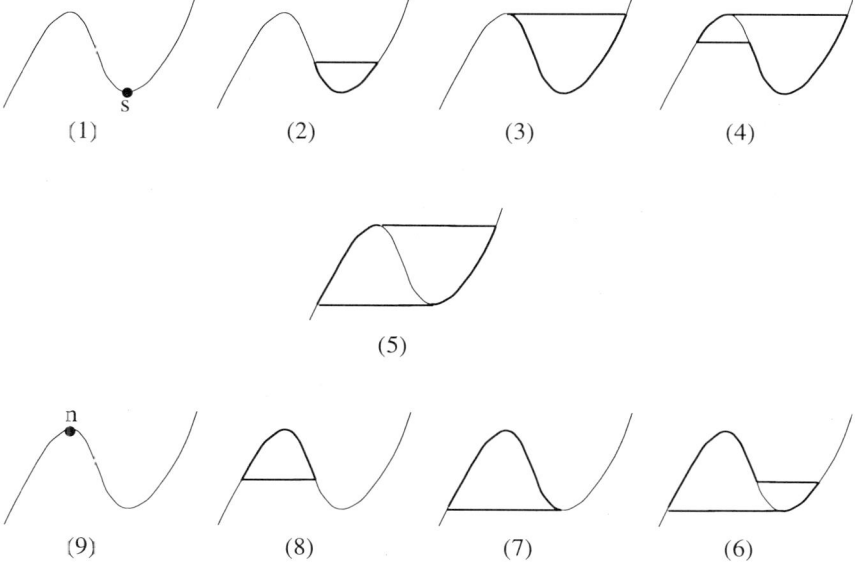

Figure 3
List of limit periodic sets

The others ((2) - (4)) are commonly called "canards" (ducks), because of the shape of the 4th case. In these cases (2) -(4) we will call them respectively l.p.s. of type I (canard sans tête (without a head)), l.p.s. of type II (canard à petite tête (with small head)) and l.p.s. of type III (canard (avec tête)).

Let us also take into account the height of the "back" or of their "chin" as in figure 4, introducing Γ_y^I and Γ_y^{III}. We denote the l.p.s. of type II by Γ^{II}

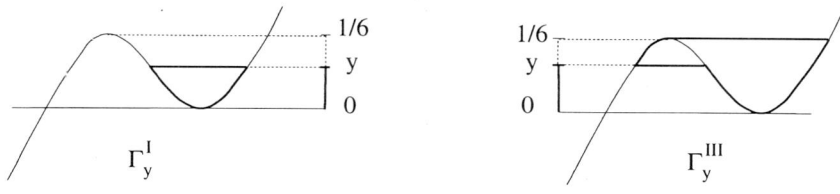

Figure 4
List of "canards"

We remark that Γ_0^I is the small l.p.s. Γ_0, Γ_0^{III} the big one Γ_B^I while $\Gamma_{1/6}^I = \Gamma_{1/6}^{III} = \Gamma^{II}$. We will hence for Γ_y^I and Γ_y^{III} only speak of a l.p.s. of canard type when $y > 0$.

The orientation of the limit periodic set is the one induced by the nearby limit cycles and is of course compatible with the flow of X_0.

Our way to describe the "canard phenomenon" is now precisely stated in the following theorem, whose proof we will give in this paper, mainly using geometric ideas.

We define

$$k(y) = \int_0^{x(y)} x(1+x)^2 dx,$$

where $x(y)$ is the largest negative solution of $\{x^2/2 + x^3/3 = y\}$ and $y \in [0, 1/6]$. The function k is C^1, $k'(y) = 1 + x(y)$, k is analytic on $]0, 1/6[$, $k(0) = 0$ and $k(1/6) = 1/12$. In terms of $x(y)$ this function is polynomial and it coincides with the one given in [BCD].

Theorem 1

There exists a curve $C_0 = \{a = c_0(\varepsilon)\}$ with $c_0(\varepsilon) = \sqrt{\varepsilon}\,\bar{a}(\sqrt{\varepsilon})$ and $\bar{a}$ a C^∞-function with $\bar{a}(0) = 0$ and $\bar{a}'(0) = -1$, such that for any continuous curve $C = \{a = c(\varepsilon)\}$ with $c(\varepsilon) \leq 0$ and $c(0) \in [-1/2, 0]$ we have (using $y \in]0, 1/6[$) :

1) $\displaystyle\lim_{\varepsilon \to 0} \Gamma_{\varepsilon, c(\varepsilon)} = \Gamma_0 \quad \Leftrightarrow \quad$ *for small $\varepsilon > 0$: $c(\varepsilon) > c_0(\varepsilon)$ and*
$$\overline{\lim}(-\varepsilon \log(c(\varepsilon) - c_0(\varepsilon))) \leq 0$$

2) $\displaystyle\lim_{\varepsilon \to 0} \Gamma_{\varepsilon, c(\varepsilon)} = \Gamma_y^I \quad \Leftrightarrow \quad$ *for small $\varepsilon > 0$: $c(\varepsilon) \geq c_0(\varepsilon)$ and*
$$\lim(-\varepsilon \log(c(\varepsilon) - c_0(\varepsilon))) = k(y)$$

3) $\displaystyle\lim_{\varepsilon \to 0} \Gamma_{\varepsilon, c(\varepsilon)} = \Gamma^{II} \quad \Leftrightarrow \quad \underline{\lim}(-\varepsilon \log |c(\varepsilon) - c_0(\varepsilon)|) \geq k(1/6)$

4) $\displaystyle\lim_{\varepsilon \to 0} \Gamma_{\varepsilon, c(\varepsilon)} = \Gamma_y^{III} \quad \Leftrightarrow \quad$ *for small $\varepsilon > 0$: $c(\varepsilon) \leq c_0(\varepsilon)$ and*
$$\lim(-\varepsilon \log(c_0(\varepsilon) - c(\varepsilon))) = k(y)$$

5) $\displaystyle\lim_{\varepsilon \to 0} \Gamma_{\varepsilon, c(\varepsilon)} = \Gamma_B \quad \Leftrightarrow \quad$ *for small $\varepsilon > 0$: $c(\varepsilon) < c_0(\varepsilon)$ and*
$$\overline{\lim}(-\varepsilon \log(c_0(\varepsilon) - c(\varepsilon))) \leq 0$$

Theorem 1 implies that in order to obtain a limit periodic set of canard type as limit of a 1-parameter family of limit cycles $\Gamma_{\varepsilon,\gamma(\varepsilon)}$ we have to choose a $\gamma(\varepsilon)$ having a flat contact with $c_0(\varepsilon)$. It also describes the exact asymptotics of the curve $(\varepsilon, \gamma(\varepsilon))$, needed to find as limit of $\Gamma_{\varepsilon,\gamma(\varepsilon)}$ a canard l.p.s. of a given type and with a given "height" $y \in\,]0, 1/6]$.

While proving theorem 1 we will also give a proof of the following result, which is clearly related to theorem 1, but not a direct consequence of it. In its statement we use the notations introduced in theorem 1.

Theorem 2

For each $\delta > 0$, there exists $\varepsilon_0(\delta) > 0$ and $K(\delta) > 0$ such that for $0 < \varepsilon < \varepsilon_0(\delta)$, one necessarily has $a \in [c_0(\varepsilon) - e^{-K(\delta)/\varepsilon}, c_0(\varepsilon) + e^{-K(\delta)/\varepsilon}]$ when one of the following conditions holds

$$d(\Gamma_{\varepsilon,a}, \Gamma_y^I) < \delta/2 \quad with \quad y \in [\delta, 1/6]$$

or

$$d(\Gamma_{\varepsilon,a}, \Gamma_y^{III}) < \delta/2 \quad with \quad y \in [\delta, 1/6].$$

Remark :

1. As a possible $K(\delta)$ we can choose any $k(\delta')$ with $\delta' \leq \delta/2$.

2. As a consequence of theorem 2 and for a $\delta/2$-resolution, one can see no other l.c. but small ones ($\delta/2$-close to a small l.p.s.) for $a \in [c_0(\varepsilon) + e^{-K(\delta)/\varepsilon}, 0]$ and big ones ($\delta/2$-close to a big l.p.s.) for $a \in [-1/2, c_0(\varepsilon) - e^{-K(\delta)/\varepsilon}]$, when $0 < \varepsilon < \varepsilon_0(\delta)$ and $\varepsilon_0(\delta)$ is taken sufficiently small.

3. Of course, if in theorem 2 we restrict the choice of $\Gamma_{\varepsilon,a}$ to a given 1 parameter curve $(\varepsilon, a(\varepsilon))$ with $a(\varepsilon) \stackrel{\varepsilon \to 0}{\to} 0$ then the conclusion holds for $\varepsilon > 0$ sufficiently small, as a consequence of theorem 1.

2 Global desingularization

2.1 Survey of the construction

Consider the family $X_{\varepsilon,a}$ given in (5) for $(\varepsilon, a) \in [0, \infty[\times [-\frac{1}{2}, \infty[= \wedge$. As it was said before, the study for any a is obtained by means of a symmetry with respect to $a = -\frac{1}{2}$.

We need to desingularize $X_{\varepsilon,a}$ because some singular points are not elementary. (i.e. have a trivial spectrum of eigenvalues). These non elementary points form the 2 lines $S = \{(x, y) = s = (0, 0), \varepsilon = 0\}$ and $N = \{(x, y) = n = (-1, \frac{1}{6}), \varepsilon = 0\}$. On S, the point $s_0 = \{s\} \times \{(0, 0)\}$ will be considered to be more degenerate, because it is a limit of singular points of $X_{\varepsilon,a}$, for $\varepsilon \neq 0$. The singular points of $X_{\varepsilon,a}$, not contained in $N \cup S$, are non-degenerate for $\varepsilon \neq 0$ and form surfaces of normally hyperbolic points for $\varepsilon = 0$.

The desingularization will consist of a sequence of weighted blow ups, i.e. locally making coordinate changes of the form $x = u^n \overline{x}$, $y = u^m \overline{y}$, $\varepsilon = u^p \overline{\varepsilon}$, $a = u^q \overline{a}$, with $\overline{x}^2 + \overline{y}^2 + \overline{\varepsilon}^2 + \overline{a}^2 = 1$ and $n, m, p, q \in I\!N$ well chosen. This method of blowing up is explained in [DeR] and some applications are presented in [R], [DR]. We refer to these papers for more explanation.

In [D] one can find an application to the classical Andronov-Hopf bifurcation, merely done for didactical purposes. Experience indeed reveals that it can take some time to get insight in what the method exactly does.

The method is a natural extension of the well known technique of rescaling, where one uses a similar kind of coordinate change, however taking $\overline{\varepsilon}^2 + \overline{a}^2 = 1$ and in general restricting $(\overline{x}, \overline{y})$ to a large ball in $I\!R^2$.

Blow up of the family adds extra information, by imposing a further study in a "chart" given by $\overline{x}^2 + \overline{y}^2 = 1$ and $(\overline{\varepsilon}, \overline{a}) \sim (0, 0)$. Both "charts" together provide a complete study along the sphere defined by $\overline{x}^2 + \overline{y}^2 + \overline{\varepsilon}^2 + \overline{a}^2 = 1$. A more detailed description of this procedure can be found in 2.2. We then also show that it is better to work along lines in parameter space given by $(\overline{\varepsilon}, \overline{a}) = (v^p E, v^q A)$ with $E^2 + A^2 = 1$. This permits to reduce a study near $S^3 \times [0, \infty[$, in the coordinates $((\overline{x}, \overline{y}, \overline{\varepsilon}, \overline{a}), u)$, to a study near $S^2 \times [0, \infty[$, in the coordinates $((\overline{x}, \overline{y}, v), u)$, however depending on a 1-dimensional cyclic parameter $(E, A) \in S^1$.

10

The fact which presumably most complicates an easy understanding of the method is that globally near the sphere $\bar{x}^2 + \bar{y}^2 + \bar{\varepsilon}^2 + \bar{a}^2 = 1$, as well as in the chart given by $\bar{x}^2 + \bar{y}^2 = 1$, the application of the blow up does not lead to a 2-parameter family of vector fields, unlike for the usual rescaling. At each step of the desingularization a geometrical object $\mathcal{E} = (M, \pi, \wedge, X)$, called a **foliated local vector field** will be produced : M and $\wedge$ are analytic manifolds, $\pi : M \to \wedge$ is a surjective analytic map and X a **local vector field**. Let us recall that a local vector field is given by a collection $\{(U_i, X_i)\}_{i \in I}$ where $\{U_i\}_{i \in I}$ is an open cover of M, while X_i is an analytic vector field on U_i. Moreover, if $U_i \cap U_j \neq \emptyset$, there exists a positive analytic function g_{ij} such that $X_i = g_{ij} X_j$ on $U_i \cap U_j$. Of course the local vector field X is defined to be the maximal collection compatible with the given one (see [DeR]). (Such a structure is often called an oriented foliation with singularities, but we prefer to reserve the term foliation for another object that we will introduce hereafter). We suppose that X is compatible with π in the sense that $d\pi(x)[X(x)] = 0$ for any $x \in M$. Next we suppose that π is locally given in coordinates by non trivial normal maps. The foliation defined by the regular fibers of π (supposed to be 2-dimensional submanifolds, diffeomorphic to $\mathbb{R}^2$) extends to a unique foliation $\mathcal{F}$, outside a singular set σ and X is tangent to $\mathcal{F}$.

We begin with $M_0 = \mathbb{R}^2 \times \wedge$, the projection π_0 of M_0 on $\wedge$ and the vector field X^0 are defined by the family $X_{\varepsilon,a}$ on M_0. Such a (genuine) vector field may be considered as a local one. (This means that we allow changing it locally by a positive multiplicative function). The **first step** will be the blow up of the point s_0 with a blow-up map as described below. This produces a new foliated local vector $\mathcal{E}_1 = (M_1, \pi_1, \wedge, X^1)$. Here M_1 is the blown-up space, $\pi_1 = \pi_0 \circ \Phi_1$, where Φ_1 is the blow up map, $\Phi_1 : M_1 \to M_0$. The local field X^1 is defined by $X^0 = X_{\varepsilon,a}$ on $M_0 \setminus \{s_0\}$ (identified to a part of M_1) and a vector field $\overline{X}_1$ on T_1, the domain of the blowing up (see 2.2 for a precise definition).

The blown up locus $\sum_1$ of the blow-up Φ_1 is diffeomorphic to a 3-disk (representing a 3-hemi-sphere, since the parameter space is restricted to $\varepsilon \geq 0$). This set $\sum_1$ intersects the singular fiber $\hat{F}_0$ along a circle σ_1, singular set of the foliation $\mathcal{F}_1$. (We have $\pi_1^{-1}(0) = \sum_1 \cup \hat{F}_0$). The leaves of $\mathcal{F}_1$ are the regular fibers $\pi_1^{-1}(\lambda) = \hat{F}_\lambda$ for $\lambda = (\varepsilon, a) \neq 0 \in \wedge$ and 2-disks filling $\sum_1 \setminus \sigma_1$, with σ_1 as common boundary.

To simplify, we continue to call N, S the closure in M_1 of the counter-images of N, S in $M_1 \backslash \pi_1^{-1}(0)$. The second blow-up is made at $n_0 = \{n\} \times (0,0) \in N \subset M_1$. We will obtain a new foliated local vector field $\mathcal{E}_2 = (M_2, \pi_2, \wedge, X^2)$. The local field X^2 is defined by $X_{\varepsilon,a}$ in $M \backslash \{s_0, n_0\}$, $\overline{X}_1$ on T_1 and $\overline{X}_2$ on T_2, domain of the second blow-up. The blowing-up locus of the new blow-up is again a 3-disk Σ_2. As above Σ_2 intersects the singular fiber $\hat{F}_0$ along a circle σ_2. Now, $\hat{F}_0$ is diffeomorphic to $I\!\!R^2$ minus 2 open disks. The foliation $\mathcal{F}_2$ has $\sigma_1 \cup \sigma_2$ as singular set and as leaves has the regular fibers of π_2, and the 2-disks in $\Sigma_2 \backslash \sigma_2$ and in $\Sigma_1 \backslash \sigma_1$.

Again, we reuse the names N and S for the closure of their respective counter-images in $M_2 \backslash \pi_2^{-1}(0)$. The following steps consist in blowing up N, S in order to obtain a local field X of which the points are all elementary. This precisely means that any singular point $x \in \overline{L}$, where L is a leaf of $\mathcal{F}$ (whose closure $\overline{L}$ is an analytic submanifold of M) is an elementary singular point of $X|\overline{L}$.

Finally we blow up $\hat{F}_0$. This last blow-up is not needed for the desingularization but will be useful for the description of the center manifolds in chapter III.

After all these blow-ups we eventually obtain a foliated local vector field $\mathcal{E} = (M, \pi, \wedge, X)$. The singular sets of each blow-up $\Sigma_1, \Sigma_2, \hat{s}_i, \hat{N}_j, \tilde{F}_0$ are points of ∂M.

The singular set of the foliation is equal to $\sigma = \tilde{\sigma}_1 \cup \tilde{\sigma}_2 \cup_i \partial \hat{N}_i \cup_j \partial \hat{S}_j$. The leaves of the corresponding foliation $\mathcal{F}$ are submanifolds of M, diffeomorphic to $I\!\!R^2$, outside ∂M. On $\partial M \backslash \sigma$ they are diffeomorphic to a disk, a disk minus one point, or a disk minus 2 points, with boundary in σ. Each set $\tilde{\sigma}_i, \tilde{\sigma}_2, \hat{N}_i, \hat{S}_j$ is diffeomorphic to $[0,1] \times S^1$ ($\tilde{\sigma}_1, \tilde{\sigma}_2$ are the result of the blowing up of σ_1, σ_2 in the last step).

A geometric picture of M is presented in figure 5. In fact the picture is meant to be a representation of M in dimension 3, with leaves of $\mathcal{F}$ pictured as 1-dimensional submanifolds. In this image the circles S^1 are replaced by the 0-sphere S^0, so that $\tilde{\sigma}_2$ for instance appears as $[0,1] \times S^0$ (union of 2 intervals).

Typical sections by 3-spaces $P_{\overline{\lambda}_0}$, Q_a will be defined later on.

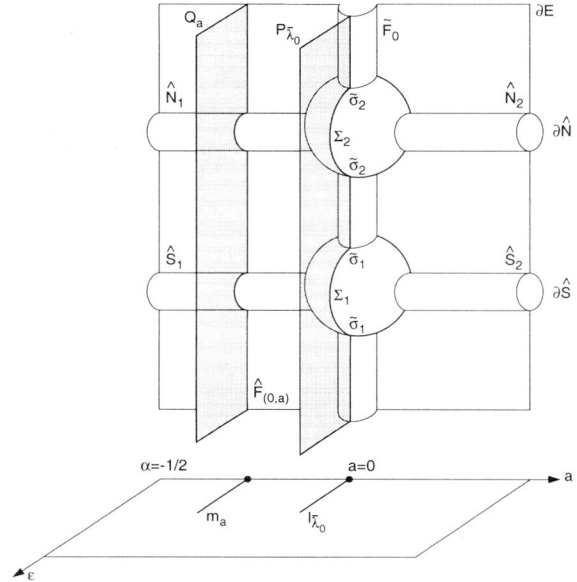

Figure 5
3-dimensional representation of M

2.2 Blow-up at $s_0 = 0 = (0, 0, 0, 0)$

Let $S_+^3 = \{(\overline{x}, \overline{y}, \overline{\varepsilon}, \overline{a}) \in I\!\!R^4 | \overline{x}^2 + \overline{y}^2 + \overline{\varepsilon}^2 + \overline{a}^2 = 1, \overline{\varepsilon} \geq 0\}$.
The blow-up ϕ_1 is the map :

$$\phi_1(\overline{x}, \overline{y}, \overline{\varepsilon}, \overline{a}, u) = (u\overline{x}, u^2\overline{y}, u^2\overline{\varepsilon}, u\overline{a}) \tag{6}$$

defined on $T_1 = S_+^3 \times [0, U[\to M_0 = I\!\!R^2 \times \wedge$.
The number $U > 0$ is chosen small enough so that $\phi_1(T_1)$ does not contain points (x, y, ε, a) with $(x, y) \in L_1$ (see figure 1).

As above, let X^0 be the vector field defined by the family (5). Since $X^0(0) = 0$, there exists a vector field $\hat{X}_1$ on T_1 with $(\phi_1)_*(\hat{X}_1) = X^0$. We can divide $\hat{X}_1$ by the function $u : \hat{X}_1 = u\overline{X}_1$, where $\overline{X}_1$ will be called the **desingularized vector field** (on T_1). Now, as explained in [DeR], we can join T_1 with $M_0 \backslash \{0\}$ by the map ϕ_1 to obtain $M_1 = ((S_+^3 \times [0, U[) \cup M_0 \backslash \{0\})/(m \sim \phi_1(m))$. Let Φ_1 be the map from M_1 to

M_0 which restricts to ϕ_1 on T_1 and to the identity on $M_0\backslash\{0\}$. The local vector field X_1 is defined by X^0 on $M_0\backslash\{0\}$ and by $\overline{X}_1$ on T_1, where $M_0\backslash\{0\}$ and T_1 are identified with open subsets of M_1. If $\pi_1 = \pi_0 \circ \Phi_1$, we have $d\pi_1(m)|X_1(m)] = 0$. The only critical value of π_1 is $\lambda = 0$ and $\pi_1^{-1}(0) = \hat{F}_0 \cup \Sigma_1$. Here $\hat{F}_0$, the singular fiber, is the result of the blow up on $F_0 = \pi_0^{-1}(0)$ and $\Sigma_1 = S_+^3 \times \{0\}$ in the domain T_1. We have $\sigma_1 = \Sigma_1 \cap \hat{F}_0 = \{(\overline{x},\overline{y},\overline{\varepsilon},\overline{a},u) \in T_1|\overline{\varepsilon} = \overline{a} = u = 0\}$: it is a large circle of S_+^3. If we define, for each $\overline{\lambda}_0 = (\overline{\varepsilon}_0, \overline{a}_0) \in S_+^1 = \{\overline{\varepsilon}_0^2 + \overline{a}_0^2 = 1, \varepsilon_0 \geq 0\}$,

$$\overline{D}_{\overline{\lambda}_0} = \{(\overline{x},\overline{y},\overline{\varepsilon},\overline{a}) \in S_+^3|\ \overline{\varepsilon}\ \overline{a}_0^2 = \overline{\varepsilon}_0\overline{a}^2\} \tag{7}$$

it is easy to see that the interiors $D_{\overline{\lambda}_0}$ of these closed 2-disks fill $\Sigma_1 \backslash \sigma_1$, with σ_1 as common boundary and together with the regular fibers $\hat{F}_\lambda = \pi_1^{-1}(\lambda)$, for all $\lambda \in \wedge\backslash\{0\}$, define a 2-dimensional foliation $\mathcal{F}_1$ on $M_1\backslash\sigma_1$. We say that σ_1 is the singular set of $\mathcal{F}_1$. Of course, X_1 is tangent to $\mathcal{F}_1$ in $M_1\backslash\sigma_1$ and tangent to σ_1 along σ_1.

To have a geometric picture of $\mathcal{F}_1$, we can construct a 3-dimensional section of M_1 as follows : let $\overline{\lambda}_0 = (\overline{\varepsilon}_0, \overline{a}_0) \in S_+^1$ and let $\ell_{\overline{\lambda}_0}$ be the following curve :

$$\ell_{\overline{\lambda}_0} = \{\lambda \in \wedge|\lambda = (u^2\overline{\varepsilon}_0, u\overline{a}_0), u > 0\} \subset \wedge \tag{8}$$

Then, we consider the section $P_{\overline{\lambda}_0}$ to be the closure in M_1 of $\pi_1^{-1}(\ell_{\overline{\lambda}_0})$. It is foliated by $\mathcal{F}_1$ with two leaves $int(\hat{F}_0)$ and $D_{\overline{\lambda}_0}$ in $\pi_1^{-1}(\{0\})$ and $\hat{F}_\lambda = \pi_1^{-1}(\lambda)$, $\lambda \in \ell_{\overline{\lambda}_0}$ as other leaves.

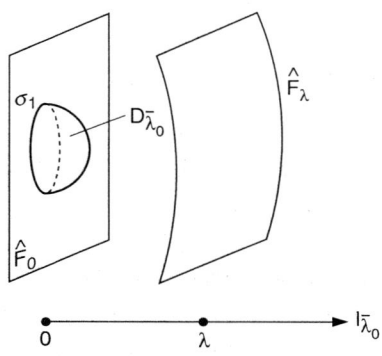

Figure 6

Foliation $\mathcal{F}_1$ in $P_{\overline{\lambda}_0} \subset M_1$.

We obtain the whole foliation $\mathcal{F}_1$ by considering all $\overline{\lambda}_0 \in S^1_+$. The closed disks $\overline{D}_{\overline{\lambda}_0}$ fill up $S^3_+ \approx \Sigma_1$.

Now we turn back to the local field X_1. Of course it suffices to describe $\overline{X}_1$ on T_1. The definition of local vector field itself allows to replace $\overline{X}_1$ by proportional vector fields on coordinate domains covering T_1. As in [DR], we will use 2 systems of charts : the **family rescaling charts** and the **phase directional rescaling charts**. A similar construction as described above (take the related $\hat{X}_1$, next divide by the local blow-up parameter u) will define the required vector fields that are proportional to $\overline{X}_1$ (in different coordinates).

2.2.1 Family Rescaling

Let $(\overline{\varepsilon}, \overline{a}) \in S^1_+$, and $(\overline{x}, \overline{y}) \in D^-$, D^- being some (large) disk in $I\!\!R^2$; $u \in [0, U[$. In this chart the desingularized vector field $\overline{X}_1$ is equivalent (up to a positive multiplicative factor) to the following family of vector fields, which we call again $\overline{X}_1$ to simplify notation (after all "it is" $\overline{X}_1$ in a different coordinate system) :

$$
\overline{X}_1 : \begin{cases} \dot{\overline{x}} = \overline{y} - \dfrac{\overline{x}^2}{2} - \dfrac{u}{3}\overline{x}^3 \\[2mm] \dot{\overline{y}} = \overline{\varepsilon}(\overline{a} - \overline{x}) \end{cases} \tag{9}
$$

In practice, we will replace this chart by two subcharts :

$$
\begin{aligned}
FR1 &: \quad \overline{a} \in [-A_0, A_0], \overline{\varepsilon} = 1 \\
FR2 &: \quad \overline{a} = \pm 1, \overline{\varepsilon} \in [0, \varepsilon_0[
\end{aligned}
$$

for a large $A_0 > 0$ and where ε_0 may be taken small. The equations of $\overline{X}_1$ in these charts are again given by (9).

Consider first $\overline{X}_1$ on the blown-up locus $\{u = 0\}$. It will be convenient to study directly the phase portrait of $\overline{X}_1$ on the disk $\overline{D}_{\overline{\lambda}_0}$ (for each $\overline{\lambda}_0 = (1, \overline{a})$ in the chart FR1 for instance; D^- is a disk in $D_{\overline{\lambda}_0} = \overline{D}_{\overline{\lambda}_0} \backslash \sigma_1$). The vector field on $\overline{D}_{\overline{\lambda}_0}$ may be considered as to be obtained by a compactification of (9), the circle at infinity being $\sigma_1 = \partial \overline{D}_{\overline{\lambda}_0}$.

The study near σ_1 will be obtained in the phase directional rescaling chart to be studied below. Along σ_1 we will find four singular points of $\overline{X}_1$, all of them hyperbolic when restricted to σ_1.

Seen on the disks $\overline{D}_{\overline{\lambda}_0}$, two of them, namely s_2 and s_3 are semi-hyperbolic saddle-type points, s_1 is a hyperbolic sink, and s_4 a hyperbolic source. At finite distance (i.e. in the given $\overline{D}$) the vector field $\overline{X}_1$, for $\overline{\lambda}_0 = (1, \overline{a})$, is given by :

$$\overline{X}_1 : \begin{cases} \dot{\overline{x}} = \overline{y} - \dfrac{\overline{x}^2}{2} \\[3mm] \dot{\overline{y}} = \overline{a} - \overline{x} \end{cases} \tag{10}$$

For $\overline{a} = 0$, $\overline{X}_1$ is symmetric with respect to the $\overline{y}$-axis. So , the singular point 0 is a center. The parabola $\{\overline{y} = \dfrac{\overline{x}^2}{2} - 1\}$ is an invariant curve. In $\overline{D}_{(1,0)}$ it is a connection at infinity between the points s_2, s_3. When $\overline{a} \neq 0$ no periodic orbits exist. This can be seen by writing $\overline{X}_1$ as $\dot{\overline{u}} = y$, $\dot{y} = -u - (a + u)y$ and observing that this family has a rotational property with respect to the parameter a. The vector field has just one singular point $(\overline{a}, \dfrac{\overline{a}^2}{2})$ which is attracting for $\overline{a} > 0$ and repelling for $\overline{a} < 0$.

The different phase portraits of $\overline{X}_1$ on $\overline{D}_{(1,\overline{a})}$ are shown in the Figure 7 :

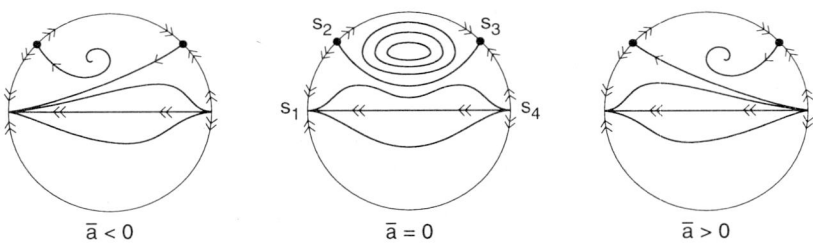

$\overline{a} < 0$ $\overline{a} = 0$ $\overline{a} > 0$

Figure 7

Phase portraits of $\overline{X}_1$ on $\overline{D}_{(1,\overline{a})}$ in FR1.

Consider now the charts FR2, again for $u = 0$. The equations of $\overline{X}_1$ are written :

$$\overline{X}_1 : \begin{cases} \dot{\overline{x}} = \overline{y} - \dfrac{\overline{x}^2}{2} \\[2mm] \dot{\overline{y}} = \overline{\varepsilon}(\pm 1 - \overline{x}) \end{cases} \tag{11}$$

For $\overline{\varepsilon} = 0$, $\overline{X}_1$ has a whole curve of singular points, namely the parabola $P = \{\overline{y} = \dfrac{\overline{x}^2}{2}\}$. Observe that, outside the top $s_1 = (0,0)$, all the points of P are normally hyperbolic. We will look closer to this point in a further desingularization.

For $\overline{\varepsilon} > 0$, $\overline{X}_1$ has a unique singular point : a repelling node at $(-1, \frac{1}{2})$ if $\overline{a} = -1$, and an attracting node at $(1, \frac{1}{2})$ for $\overline{a} = 1$; the phase portraits are the limit situations of the above ones in the FR1-chart (for $\overline{a} \neq 0$).

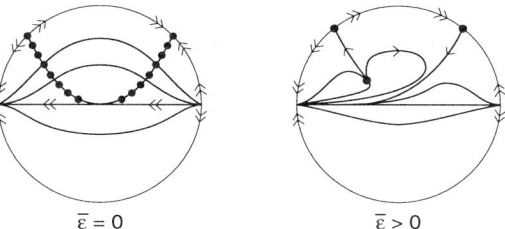

$$\overline{\varepsilon} = 0 \qquad\qquad\qquad \overline{\varepsilon} > 0$$

Figure 8
Phase portraits for $\overline{X}_1$ on $\overline{D}_{(\overline{\varepsilon}, -1)}$ in FR2.

Information about $\overline{X}_1$ for $u \neq 0$ (outside the blown up locus) will be obtained in the next part.

2.2.2 Phase-directional rescaling

Let now V be a neighbourhood of 0 in $(\overline{\varepsilon}, \overline{a})$-space. (We can choose a small neighbourhood, if the disk D^- in the previous family rescaling chart is large enough). We take $(\overline{x}, \overline{y}) \in S^1$, $(\overline{\varepsilon}, \overline{a}) \in V$ and $u \in [0, U[$. In this chart, $\overline{X}_1$ is definitely not a family of planar vector fields. However the first integrals

$$u\overline{a} = a, \, u^2 \overline{\varepsilon} = \varepsilon \tag{12}$$

induce the following relations :

$$\frac{\dot{\overline{a}}}{\overline{a}} = \frac{1}{2}\frac{\dot{\overline{\varepsilon}}}{\overline{\varepsilon}} = -\frac{\dot{u}}{u} \tag{13}$$

In order to simplify the calculations, we are not going to work with $S^1 = \{\overline{x}^2 + \overline{y}^2 = 1\}$ but with $S^1 = \{\overline{x}^4 + 2\overline{y}^2 = 1\}$ like in [DR]. This analytic circle is parametrized by

$$\overline{x} = Cs\theta \qquad\qquad \overline{y} = Sn\theta \tag{14}$$

where the analytic functions Cs, Sn are solutions of the Cauchy problem

$$\begin{cases} \dfrac{d}{d\theta}Cs\theta = -Sn\theta \quad Cs(0) = 1 \\[3mm] \dfrac{d}{d\theta}Sn\theta = Cs^3\theta \quad Sn(0) = 0 \end{cases} \tag{15}$$

They satisfy : $2Sn^2\theta + Cs^4\theta = 1$ and are T-periodic for

$$T = \sqrt{2}\int_0^1 (1-t)^{-1/2}t^{-3/4}dt \tag{16}$$

Moreover :

$$\begin{array}{ll} Cs(-\theta) = Cs(\theta) & Sn(-\theta) = -Sn(\theta) \\ Cs(T/2 - \theta) = -Cs(\theta) & Sn(T/2 - \theta) = Sn(\theta) \\ Cs(T/2 + \theta) = -Cs(\theta) & Sn(T/2 + \theta) = -Sn(\theta) \end{array} \tag{17}$$

To simplify the reading we will write C, S instead of $Cs(\theta)$ and $Sn(\theta)$. We will also only consider $\dot{\theta} = \dfrac{d\theta}{dt}$ and $\dot{u} = \dfrac{du}{dt}$ since $\dot{\overline{a}}$ and $\dot{\overline{\varepsilon}}$ are then given by (13). We get :

$$\overline{X}_1 : \begin{cases} \dot{u} = (SC^3 - \dfrac{C^5}{2} - \dfrac{1}{3}uC^6 + \overline{\varepsilon}(S\overline{a} - SC))u \\[3mm] \dot{\theta} = (-2S^2 + SC^2 + \dfrac{2}{3}uSC^3 + \overline{\varepsilon}(C\overline{a} - C^2)) \end{cases} \tag{18}$$

The singular points on $\sigma_1 = \{\overline{a} = \overline{\varepsilon} = 0\}$, parametrized by θ, are solutions of :

$$S(-2S + C^2) = 0 \tag{19}$$

We find : $\theta_1 = T/2$, $\theta_4 = 0$ (for $S = 0$), and the solutions of

$$2S - C^2 = 0 \tag{20}$$

Since $S > 0$, these solutions belong to $]0, T/2[$. The equation (20) gives $S = \dfrac{1}{\sqrt{6}}$, which corresponds to two solutions : θ_2, θ_3 with $0 < \theta_3 < T/4$ and $\theta_2 = T/2 - \theta_3$. These 4 solutions θ_i, $i = 1, \dots, 4$ are the θ-coordinates of the points s_i. Because they are simple, the singular points s_i are hyperbolic for $\overline{X}_1|\sigma_1$.

Consider now $\overline{X}_1|\hat{F}_0$ where $\hat{F}_0 = \{\bar{a} = \bar{\varepsilon} = 0\} \approx S^1 \times I\!R^+$.

The radial eigenvalue of $\overline{X}_1$ is equal to $\dfrac{1}{2}$ at s_1, $-\dfrac{1}{2}$ at s_4 and 0 at s_2, s_3 (which agrees with the fact that s_2, s_3 are the end points of the lines of singular points $L_i^- = \phi_1^{-1}(L_i)$, $i = 2, 3$). Observe that these lines are now desingularized in the sense that s_2, s_3 are normally hyperbolic in $\hat{F}_0$ and on each disk $\overline{D}_{\overline{\lambda}_0}$. The properties of $\overline{X}_1$ on $\hat{F}_0$ are summarized in the figure 9 :

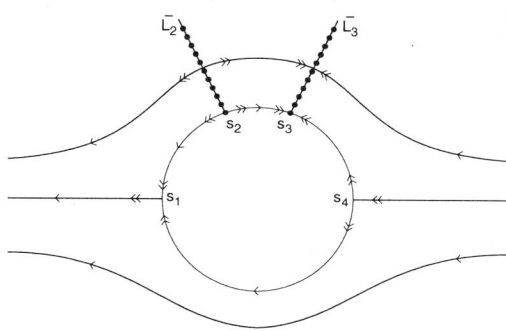

Figure 9
$\overline{X}_1$ on $\hat{F}_0$ near σ_1.

To obtain the equations for $\overline{X}_1$, we have to complete (18) by adding the equations (13) for $\bar{a}, \bar{\varepsilon}$. However we can reduce this system to a family of 3-dimensional equations. Using (13) we observe that $\bar{a}^2/\bar{\varepsilon}$ stays constant. In fact each constant corresponds to the 3-dimensional space $P_{\overline{\lambda}_0}$ as introduced above (see figure 6), for $\overline{\lambda}_0 = (\bar{\varepsilon}_0, \bar{a}_0) \in S^1_+$ with $\bar{a}_0^2/\bar{\varepsilon}_0$ as a constant $(\ell_{\overline{\lambda}_0} = \{\bar{a}^2/\bar{\varepsilon} = \bar{a}_0^2/\bar{\varepsilon}_0\})$.

We write this constant

$$\frac{\bar{a}_0^2}{\bar{\varepsilon}} = A^2 = \frac{1}{E} \qquad \text{with} \qquad A = \frac{\bar{a}_0}{\sqrt{\bar{\varepsilon}_0}} \tag{21}$$

These parameters A, E are related to the parameters $\bar{a}$, $\bar{\varepsilon}$ used in the charts FR1, FR2 : $A = \bar{a}$ in the chart FR1 and $E = \bar{\varepsilon}$ in the chart FR2. Of course, we have to remember that $\bar{a}, \bar{\varepsilon}$ change their meaning when we pass from family rescaling to phase-directional rescaling : they are parameters in the first case and variables in the second case. So, we have chosen the new names A, E to avoid confusion.

Now we introduce a third variable v, related to $\bar{\varepsilon}, \bar{a}, A$ or E by :

$$\begin{cases} \bar{\varepsilon} = v^2 \\ \bar{a} = Av \end{cases} \tag{22}$$

for $A \in [-A_0, A_0]$, A_0 large enough, or by :

$$\begin{cases} \bar{\varepsilon} = Ev^2 \\ \bar{a} = \pm v \end{cases} \tag{23}$$

for $E \in [0, E_0[$, and a small $E_0 > 0$. Note that (v, θ) are variables on $\overline{D}_{\bar{\lambda}_0}$ near $\sigma_1 = \partial \overline{D}_{\bar{\lambda}_0}$. Near σ_1, $P_{\bar{\lambda}_0}$ is now parametrized by (v, θ, u) and the equations for $\overline{X}_1$ in it are written (A, E related to $\bar{\lambda}_0$ by (21)) :

In a chart PR1 :

$$PR1 : \begin{cases} \dot{u} = U_1(\theta, v, u, A)u \\ \dot{\theta} = V_1(\theta, v, u, A) \\ \dot{v} = -U_1(\theta, v, u, A)v \end{cases} \tag{24}$$

with

$$\begin{cases} U_1 = SC^3 - \dfrac{C^5}{2} - \dfrac{1}{3}uC^6 + v^2(Av - C)S \\[2mm] V_1 = -2S^2 + SC^2 + \dfrac{2}{3}uSC^3 + v^2(Av - C)C \end{cases}$$

and in a chart PR2 :

$$PR2 : \begin{cases} \dot{u} = U_2(\theta, v, u, E)u \\ \dot{\theta} = V_2(\theta, v, u, E) \\ \dot{v} = -U_2(\theta, v, u, E)v \end{cases} \tag{25}$$

with

$$\begin{cases} U_2 = SC^3 - \dfrac{C^5}{5} - \dfrac{1}{3}uC^6 + Ev^2(\pm v - C)S \\[2mm] V_2 = -2S^2 + SC^2 + \dfrac{2}{3}uSC^3 + Ev^2(\pm v - C)C \end{cases}$$

Using these equations, we obtain for $(u = 0)$ the behaviour of $\overline{X}_1$ at infinity on the disks $D_{\overline{\lambda}_0}$. Observe also that uv is constant along trajectories of $\overline{X}_1$. These surfaces $\{uv = u_0 v_0\}$ are precisely the equations of the leaves in $P_{\overline{\lambda}_0}$, near σ_1.

To study the behaviour of $\overline{X}_1$ on $P_{\overline{\lambda}_0}$, near the points s_2, s_3 we prefer to replace the circle S^1 for $(\overline{x}, \overline{y})$ by a directional blow-up chart in the $\overline{y}$-direction, hence by taking $\overline{y} = 1$, $\overline{x} \in I\!R$. Such a chart covers the segment $]s_1, s_4[$ of σ_1 containing s_2, s_3. The blow-up formulas read :

$$x = u\overline{x}, y = u^2, \varepsilon = u^2\overline{\varepsilon}, a = u\overline{a} \tag{26}$$

The equations for $\overline{X}_1$ in the $(\overline{x}, u, \overline{\varepsilon}, \overline{a})$-coordinates are again given by (13) and :

$$\begin{cases} \dot{\overline{x}} = 1 - \frac{1}{2}2\overline{x}^2 - \frac{1}{3}u\overline{x}^3 - \frac{1}{2}\overline{\varepsilon x}(\overline{a} - \overline{x}) \\[2mm] \dot{u} = \frac{\overline{\varepsilon}}{2}(\overline{a} - \overline{x})u \end{cases} \tag{27}$$

As above, we introduce the parameters A, E and the variable v, along $\overline{x}, u$, to obtain equations for $\overline{X}_1|P_{\overline{\lambda}_0}$. We again obtain two types of charts PR1, PR2, parametrized by A or E where $\overline{X}_1$ reads :

$$PR1 : \begin{cases} \dot{\overline{x}} = 1 - \frac{1}{2}\overline{x}^2 - \frac{1}{3}u\overline{x}^3 - \frac{1}{2}v^2(Av - \overline{x})\overline{x} \\[2mm] \dot{u} = \frac{v^2}{2}(Av - \overline{x})u \\[2mm] \dot{v} = -\frac{v^2}{2}(Av - \overline{x})v \end{cases} \tag{28}$$

with $A \in [-A_0, A_0]$.

$$PR2 : \begin{cases} \dot{\overline{x}} = 1 - \frac{1}{2}\overline{x}^2 - \frac{1}{3}u\overline{x}^3 - \frac{1}{2}Ev^2(\pm v - \overline{x})\overline{x} \\[2mm] \dot{u} = \frac{Ev^2}{2}(\pm v - \overline{x})u \\[2mm] \dot{v} = -\frac{Ev^2}{2}(\pm v - \overline{x})v \end{cases} \tag{29}$$

with $E \in [0, E_0[$.

The coordinates of s_2, s_3 respectively are :

$$\begin{cases} s_2 = (\overline{x} = -\sqrt{2}, u = 0, v = 0) \\ s_3 = (\overline{x} = \sqrt{2}, u = 0, v = 0) \end{cases} \tag{30}$$

The phase portrait for $\overline{X}_1$ in these charts and for $\{uv = 0\}$ is presented in figure 10.

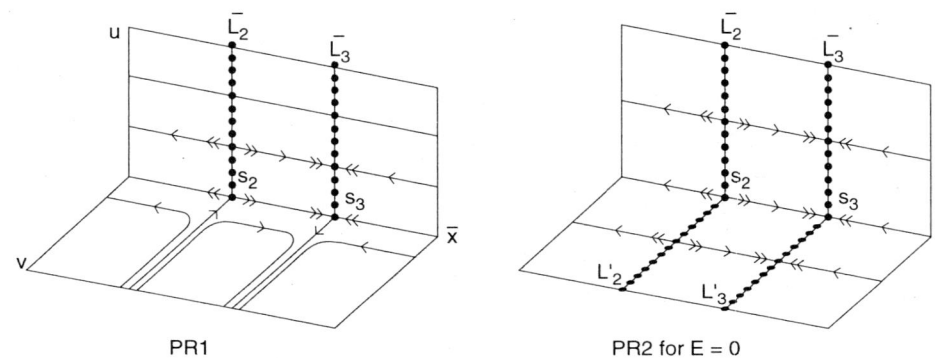

Figure 10
Phase portraits for $\overline{X}_1|\{uv = 0\}$ in PR-charts.

2.2.3   Global pictures for X^1 near s_0

It is possible to glue together the different images obtained in 2.2.1 and 2.2.2 in some global picture for X^1. Recall that we have just blown-up one point to obtain a new space M_1 with a local field X^1. Next the 4-dimensional space M_1 is cut by 3-spaces $P_{\overline{\lambda}_0}$. A 3-dimensional representation of this is presented in figure 11, where the fibers are reduced to the dimension 1 :

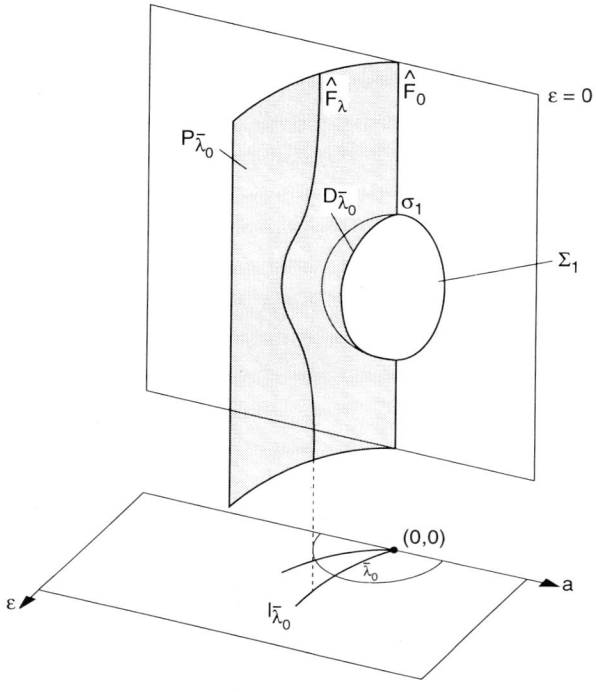

Figure 11

3-dimensional representation of $\mathcal{E}_1$.

In the next figure, we present phase portraits in different spaces $P_{\overline{\lambda}_0}$; restricted to a neighbourhood of $\overline{D}_{\overline{\lambda}_0}$.

The local vector field X^1 on M_1 has 2 lines of non-hyperbolic singular points, projecting on N, S and which we again call N, S. The first one is contained in the part $M_0 \backslash \{s_0\}$ and may in fact be identified with N. The second one has 2 connected components which we call S_1, S_2; they respectively end at the points $s_{11} = (\overline{\varepsilon} = 0, \overline{a} = 1, \overline{x} = \overline{y} = 0)$ and $s_{12} = (\overline{\varepsilon} = 0, \overline{a} = -1, \overline{x} = \overline{y} = 0)$ in the respective charts FR1 and FR2 (e.g. see s_{11} in figure 8). We can look at them in the 3-space of M_1, equal to $P_{(0,1)} \cup P_{(0,-1)}$, as we do in figure 13.

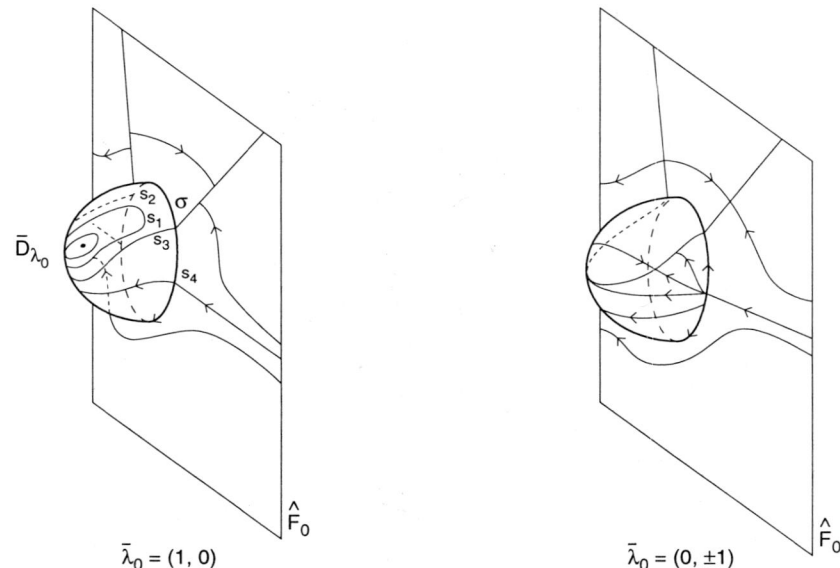

$$\bar{\lambda}_0 = (1, 0) \qquad\qquad\qquad \bar{\lambda}_0 = (0, \pm 1)$$

Figure 12

Phase portraits of $X_-^1 | P_{\bar{\lambda}_0}$ near $\overline{D}_{\bar{\lambda}_0}$

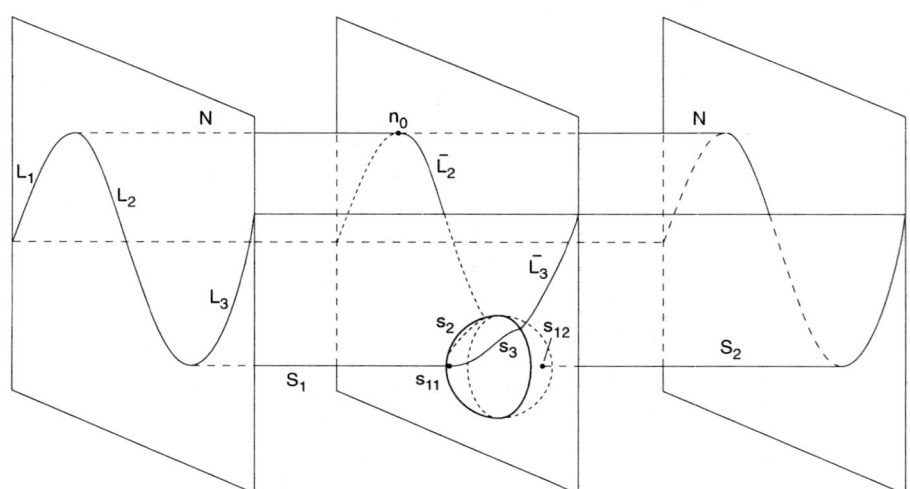

Figure 13

The lines N, S in $P_{(0,1)} \cup P_{(0,-1)} \subset M_1$.

The type of the local vector field X_1 is constant along the lines N, S_1, S_2. To finish the desingularization, in the sense of [DeR] it would be sufficient to blow up X_1 along these 3 lines. But in order to obtain a simpler description of the center manifolds in the next part we will need further blow ups. We first blow up the point n_0 in a way to be coherent with the blow-up of s_0.

2.3   Blow up at $n_0 = (-1, 1/6, 0, 0)$

As we observed before $M_0 \backslash s_0$ is identified to an open subset of M_1, so that we can use the initial coordinates (x, y, ε, a) around n_0 and represent X_1 by X.

We want to blow up at the point $n_0 = (-1, 1/6, 0, 0)$. For this we introduce local coordinates $\underline{x}, \underline{y}$ as :

$$x = -1 + \underline{x} \quad . \quad\quad y = \frac{1}{6} + \underline{y} \tag{31}$$

The equations of $X_{\varepsilon,a}$ in these coordinates are given by :

$$\begin{cases} \dot{\underline{x}} = \underline{y} + \dfrac{1}{2}\underline{x}^2 - \dfrac{1}{3}\underline{x}^3 \\[2mm] \dot{\underline{y}} = \varepsilon(1 + a - \underline{x}) \end{cases} \tag{32}$$

The blow up map will be defined by the formulas :

$$\underline{x} = u'^2 x', \underline{y} = u'^4 y', \varepsilon = u'^6 \varepsilon', a = u'^3 a' \tag{33}$$

Let again $S_+^3 = \{x'^2 + y'^2 + \varepsilon'^2 + a'^2 = 1, \varepsilon' \geq 0\}$.

Formulas (33) give a blow up map ϕ_2 :

$$\phi_2 : T_2 = S_+^3 \times [0, U'[\rightarrow M_0 \backslash \{s_0\} \subset M_1.$$

The numbers U, U' are chosen small enough so that $\overline{\phi_2(T_2)}$ does not intersect $\overline{\phi_1(T_1)}$. We can define a vector field $\overline{X}_2$ on T_2 as $\overline{X}_2 = \dfrac{1}{u'^2}\hat{X}_2$ where $\hat{X}_2$ is given by $(\phi_2)_*(\hat{X}_2) = X^0$. This blow up is quite similar to the one at s_0, and is compatible with it in the sense that one can go from one to the other with an analytic coordinate change which in parameter space is given by $u = u'^3, (\overline{a}, \overline{\varepsilon}) = (a', \varepsilon')$; remark that $u \cdot u' \neq 0$ outside $\{n_0, s_0\}$. In particular we can use ϕ_2 to construct a second blown up space $M_2 =$

$T_2 \cup (M_1 \backslash \{n_0\})/(m \sim \phi_2(m))$. Let $\Phi_2 : M_2 \to M_1$ be the corresponding blow-up map extending $\phi_2 | T_2$. We call X^2 the new local field equal to $\overline{X}_2$ on T_2 and to X^1 on $M_1 \backslash \{n_0\}$ (i.e. equal to X^0 on $M_0 \backslash \{s_0, n_0\}$ and to $\overline{X}_1$ on T_1).

Let $\Sigma_2 = S_+^3 \times \{0\} \subset T_2$ be the blown-up locus of ϕ_2 and $\sigma_2 = \Sigma_2 \cap \hat{F}_0$. Let $\pi_2 = \pi \circ \Phi_1 \circ \Phi_2$. For simplicity in notation we still keep $\hat{F}_\lambda$ to denote the fibers of π_2; we will also not change the notation of the $P_{\overline{\lambda}_0}$. This convention will be kept when performing more blow-ups. Now $\hat{F}_0$ is diffeomorphic to $\mathbb{R}^2$ minus 2 open disks, and $\partial \hat{F}_0 = \sigma_1 \cup \sigma_2$. As above, let $P_{\overline{\lambda}_0}$ be the closure in M_2 of $\pi_2^{-1}(\ell_{\overline{\lambda}_0})$. (It is now the counter image by Φ_2 of the $P_{\overline{\lambda}_0}$ defined in 2.2.).

The equation of $P_{\overline{\lambda}_0}$ in the chart T_2 is

$$P_{\overline{\lambda}_0} \cap T_2 = \{(x', y', \varepsilon', a', u') | \varepsilon' \overline{a}_0^2 = \overline{\varepsilon}_0 a'^2\} \tag{34}$$

So, $P_{\overline{\lambda}_0}$ is a sub-manifold (with corners) of M_2. Its intersection with Σ_2 is the 2-disk :

$$\overline{D}'_{\overline{\lambda}_0} = \{(x', y', \varepsilon', a') \in S_+^3 | \varepsilon' \overline{a}_0^2 = \overline{\varepsilon}_0 a'^2\} \tag{35}$$

These 2-disks fill Σ_2. Their interiors $D'_{\overline{\lambda}_0}$ are leaves of $\mathcal{F}_2$. The other leaves are the regular fibers of π_2 and the disks $D_{\overline{\lambda}_0}$ defined in 2.2.

As above we cover T_2 by several charts : family rescaling and phase-directional rescaling charts.

2.3.1　Family Rescaling

Take $(\varepsilon', a') \in S_+^1 = \{(\varepsilon', a') | \varepsilon'^2 + a'^2 = 1, \varepsilon' \geq 0\}$, $(x', y') \in D'$ a large disk; $u' \in [0, U'[$. In this chart, $\overline{X}_2$ is equivalent to a family of vector fields given by :

$$\overline{X}_2 : \begin{cases} \dot{x}' = y' + \dfrac{1}{2} x'^2 - \dfrac{u'^2}{3} x'^3 \\[2mm] \dot{y}' = \varepsilon'(1 + u'^3 a' - u'^2 x') \end{cases} \tag{36}$$

Again, we replace (36) by families in to subcharts :

$$FR'1 \ : \ a' \in [-A_0', A_0'], \varepsilon' = 1$$
$$FR'2 \ : \ a' = \pm 1, \varepsilon' \in [0, \varepsilon_0'[$$

for some large A_0 and a small $\varepsilon'_0 > 0$.

We look at $\overline{X}_2$ on the blown-up locus $\Sigma_2 = \{u' = 0\}$. Again, for any $\overline{\lambda}_0 \in S^1_+$ we will draw the phase portrait of $\overline{X}_2$ on $\overline{D}'_{\overline{\lambda}_0}$, where $\partial \overline{D}'_{\overline{\lambda}_0} = \sigma_2$. The study near σ_2 will be obtained in the phase-directional rescaling chart below. At finite distance (i.e. on D') and in the chart FR'1 equation (36) has no singular points. In the charts FR'2 and for $\varepsilon' = 0$, $a' = \pm 1$, we have a curve of singular points, namely the parabola $\{y' = -\frac{1}{2}x'^2\}$. The family $\overline{X}_2$ for small u', for ε' near 0, and near $\{x' = 0, y' = 0\}$ is similar to the initial family near n_0 : this second blow-up is of no use for the desingularization.

In figure 14, we draw phase portraits for $\overline{X}_2$:

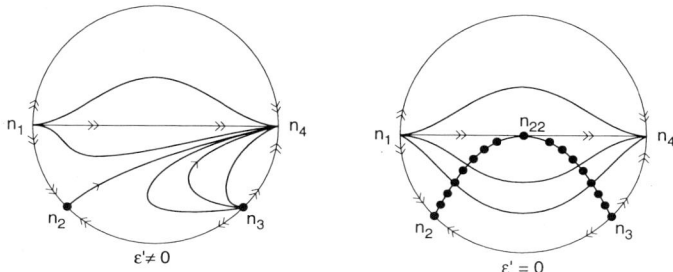

Figure 14
Phase portraits of $\overline{X}_2$ on $\overline{D}'_{\overline{\lambda}_0}$

2.3.2 Phase-directional charts

Let V' be a neighborhood of 0 in the (ε', a')-space. We take $(x', y') \in S^1$, as in 2.2.2, parametrized by $x' = Cs\theta', y' = Sn\theta'$.

The first integrals :

$$a = u'^3 a', \varepsilon = u'^6 \varepsilon' \tag{37}$$

induce the following equations :

$$\frac{\dot{a}'}{a'} = \frac{1}{2}\frac{\dot{\varepsilon}'}{\varepsilon'} = -3\frac{\dot{u}'}{u'} \tag{38}$$

Again writing C, S for $Cs\theta'$, $Sn\theta'$, we have the following equations in u', θ' in addition to (38) :

$$\overline{X}_2 : \begin{cases} \dot{u}' = \dfrac{1}{2}[C^3(S + \dfrac{C^2}{2}) + \varepsilon'S + u'^2(-\dfrac{C^6}{3} + \varepsilon'S(u'a' - C))]u' \\[4mm] \dot{\theta}' = [-S(2S + C^2) + \varepsilon'C + u'^2(\dfrac{2}{3}SC^3 + \varepsilon'C(u'a' - C))] \end{cases} \tag{39}$$

The singular points are on $\sigma_2 = \{a' = \varepsilon' = 0\}$. We find : $n_1 = \{\theta'_0 = T/2\}, n_4 = \{\theta'_4 = 0\}, n_2 = \{\theta'_1 = -\theta_2\}$ and $n_3 = \{\theta'_2 = -\theta_1\}$.

Consider $\overline{X}_2|\hat{F}_0$, where $\hat{F}_0 \cap T_2 = \{a' = \varepsilon' = 0\} \cong S^1 \times \mathbb{R}^+$.

The radial eigenvalues of $\overline{X}_2$ are : $-1/4$ at n_1, $1/4$ at n_4 and 0 at n_2, n_3, which correspond to end points of the lines $L_i^- = \Phi_2^{-1}(L_i)$, $i = 1, 2$.

The phase portrait of $\overline{X}_2$ on $\hat{F}_0 \cap T_2$ is shown in Figure 15 :

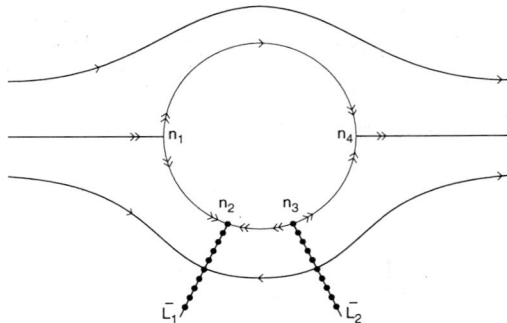

Figure 15
Phase portrait of $\overline{X}_2|\hat{F}_0 \cap T_2$

Now, as for the first rescaling we reduce the system to a family of 3-dimensional equations, by restricting it to the spaces $P_{\overline{\lambda}_0}$. We introduce the parameters A, E as in (21) in function of $\overline{\lambda}_0$. Then

$$a'^2/\varepsilon' = A^2 = \frac{1}{E} \text{ with } A = \frac{a'}{\sqrt{\varepsilon'}} \tag{40}$$

Using this relation (40), we can replace ε', a' by just one variable v' (parametrizing $D'_{\overline{\lambda}_0}$ transversally near σ_2) :

$$\begin{cases} \varepsilon' = v'^2 \\ a' = Av' \end{cases} \tag{41}$$

for $A \in [-A_0, A_0]$, and also :

$$\begin{cases} \varepsilon' = E v'^2 \\ a' = \pm v' \end{cases} \qquad (42)$$

for $E \in [0, E_0[$.

Now the equations of $\overline{X}_2$ in restriction to each $P_{\overline{\lambda}_0}$ and near σ_2 are given by the A- and E-families :

$$PR'1 : \begin{cases} \dot{u}' = U'_1(\theta', v', u', A)u' \\ \dot{\theta}' = V'_1(\theta', v', u', A) \\ \dot{v}' = -3U'_1(\theta', v', u', A)v' \end{cases} \qquad (43)$$

with

$$\begin{cases} U'_1 = \dfrac{1}{2}[C^3(S + \dfrac{C^2}{2}) + v'^2 S + u'^2(-\dfrac{C^6}{3} + v'^2 S(Au'v' - C))] \\[2ex] V'_1 = [-S(2S + C^2) + v'^2 C + u'^2(\dfrac{2}{3}SC^3 + v'^2 C(Au'v' - C))] \end{cases}$$

and

$$PR'2 : \begin{cases} \dot{u}' = U'_2(\theta', v', u', A)u' \\ \dot{\theta}' = V'_2(\theta', v', u', A) \\ \dot{v}' = -3U'_2(\theta', v', u', A)v' \end{cases} \qquad (44)$$

with :

$$\begin{cases} U'_2 = \dfrac{1}{2}[C^3(S + \dfrac{C^2}{2}) + E v'^2 S + u'^2(-\dfrac{C^6}{3} + E v'^2 S(\pm u'v' - C))] \\[2ex] V'_2 = [-S(2S + C^2) + E v'^2 C + u'^2(\dfrac{2}{3}SC^3 + E v'^2 C(\pm u'v' - C))] \end{cases}$$

These equation, for $u' = 0$, give the behaviour at infinity on the disks $D'_{\overline{\lambda}_0}$ (i.e. near $\partial \overline{D}_{\overline{\lambda}_0}$). The surfaces $\{u'^3 v' = u_0'^3 v_0'\}$ are the leaves of $\mathcal{F}_2$ in $P_{\overline{\lambda}_0}$, near σ_2.

Tot study $\overline{X}_2$ on $P_{\overline{\lambda}_0}$, near the points n_2, n_3 we prefer to use a directional blow-up in the negative $\overline{y}$-direction. It is given by :

$$\underline{x} = u'^2 x', \underline{y} = -u'^4, \varepsilon = u'^6 \varepsilon', a = u'^3 a' \qquad (45)$$

The equations of $\overline{X}_2$ in the $(x', u', \varepsilon', a')$-coordinates are again given by (38) and :

$$\begin{cases} \dot{x}' = -1 + \dfrac{x'^2}{2} - \dfrac{u'^2}{3}x'^3 + \dfrac{1}{2}x'\varepsilon'(1 - u'^2x' + u'^3a') \\[3mm] \dot{u}' = -\dfrac{1}{4}\varepsilon'(1 - u'^2x' + u'^3a')u' \end{cases} \qquad (46)$$

As above, we deduce for it equations of $\overline{X}_2$ in restriction to $P_{\overline{\lambda}_0}$. We obtain the following two families, with parameter A, E :

$$PR'1 : \begin{cases} \dot{x}' = -1 + \dfrac{x'^2}{2} - \dfrac{u'^2}{3}x'^3 + \dfrac{1}{2}x'v'^2(1 - u'^2x' + Au'^3v') \\[3mm] \dot{u}' = -\dfrac{1}{4}v'^2(1 - u'^2x' + Au'^3v')u' \\[3mm] \dot{v}' = +\dfrac{3}{4}v'^2(1 - u'^2x' + Au'^3v')v' \end{cases} \qquad (47)$$

and :

$$PR'2 : \begin{cases} \dot{x}' = -1 + \dfrac{x'^2}{2} - \dfrac{u'^2}{3}x'^3 + \dfrac{E}{2}x'v'^2(1 - u'^2x' \pm u'^3v') \\[3mm] \dot{u}' = -\dfrac{1}{4}Ev'^2(1 - u'^2x' \pm u'^3v')u' \\[3mm] \dot{v}' = +\dfrac{3}{4}Ev'^2(1 - u'^2x' \pm u'^3v')v' \end{cases} \qquad (48)$$

2.3.3   Global pictures for X^2 near $L_2 \cup \{s_0, n_0\}$

The different vector fields described above merge into the local vector field defined in M_2. We are particularly interested in the picture of X^2 restricted to the spaces $P_{\overline{\lambda}_0} = \overline{\pi_2^{-1}(\ell_{\overline{\lambda}_0})}$. Each $P_{\overline{\lambda}_0}$ is a 3-dimensional manifold with boundary and corners (along σ_1, σ_2). A 3-dimensional representation of this (like we did in figure 11 for $\mathcal{E}_1$) is represented in figure 16 :

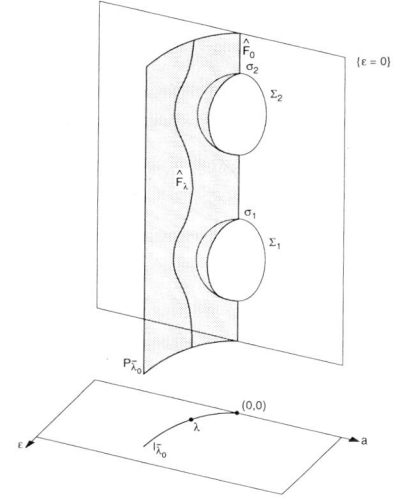

Figure 16
3-dimensional representation of $\mathcal{E}_2$.

In the next figure, we present phase portraits in the spaces $P_{\bar{\lambda}_0}$

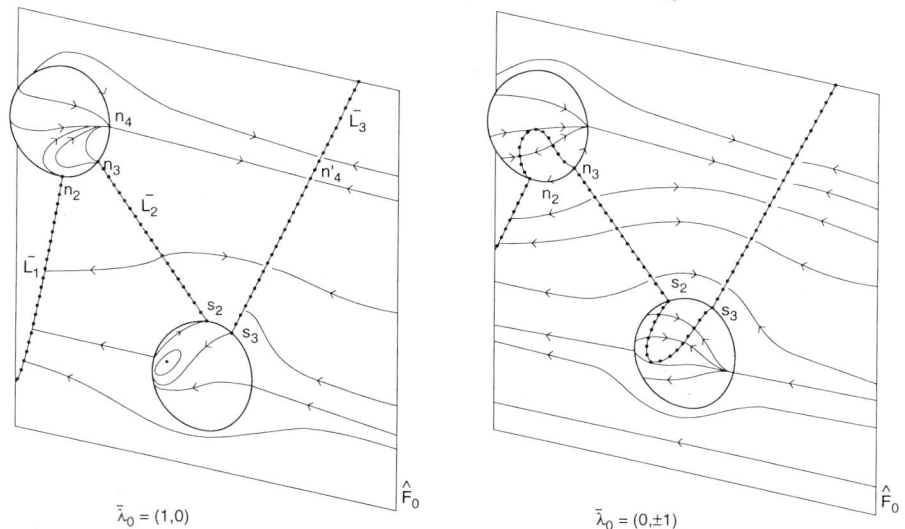

Figure 17
Phase portraits of $X^2|P_{\bar{\lambda}_0}$.

2.4   Blowing up along N, S

We continue to call by N, S their counter images in M_2.

Now in M_2, N and S above 2 connected components : S_1, S_2 for S and N_1, N_2 ending respectively at the points n_{21}, n_{22}.

Let us draw them (see figure 18), in $P_{(0,1)} \cup P_{(0,-1)}$:

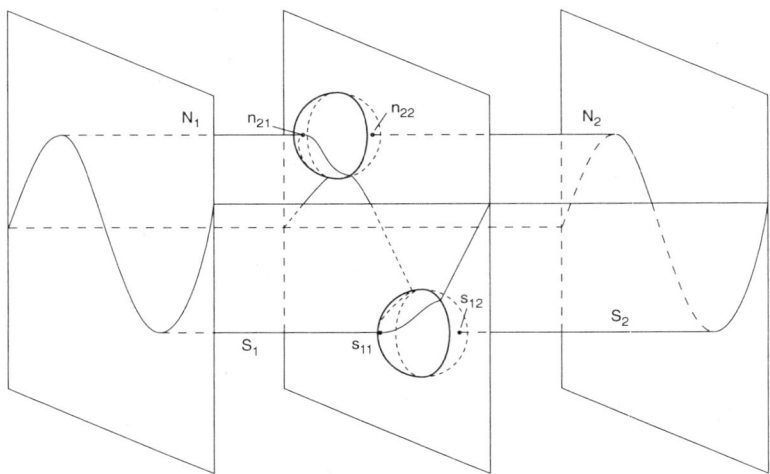

Figure 18

The lines N, S in $P_{(0,1)} \cup P_{(0,-1)} \subset M_2$.

The situation is completely similar along each of the four segments S_1, S_2, N_1, N_2. So, we limit ourselves to describe the blow up along S_1.

First, observe that X^2 along S_1 is given by 2 vector fields, namely X^0 in the chart $E_0 \backslash \{s_0, n_0\}$ and $\overline{X}_1$ near s_{11} in the chart FR2 (with $\overline{a} = -1$).

In the first chart with coordinates (x, y, ε, a), $S_1 = \{x = y = \varepsilon = 0\}$. In the second chart with coordinates $(\overline{x}, \overline{y}, \overline{\varepsilon}, u)$, $S_1 = \{\overline{x} = \overline{y} = \overline{\varepsilon} = 0\}$.

The equations for $X, \overline{X}_1$ are

$$
X : \begin{cases} \dot{x} = y - \dfrac{x^2}{2} - \dfrac{x^3}{3} \\[2mm] \dot{y} = \varepsilon(a - x) \\[2mm] \dot{\varepsilon} = 0, \dot{a} = 0 \end{cases} \quad \text{and} \quad \overline{X}_1 : \begin{cases} \dot{\overline{x}} = \overline{y} - \dfrac{\overline{x}^2}{2} - \dfrac{u}{3}\overline{x}^3 \\[2mm] \dot{\overline{y}} = \overline{\varepsilon}(-1 - \overline{x}) \\[2mm] \dot{\overline{\varepsilon}} = 0, \dot{u} = 0 \end{cases} \qquad (49)
$$

Recall that the change of coordinates is given by :

$$
x = u\overline{x}, y = u^2\overline{y}, a = -u, \varepsilon = u^2\overline{\varepsilon} \qquad (50)
$$

In the chart of X^0, we can restrict to some $\{a \leq a_0\}$ with $a_0 < 0$, so that we can limit the coordinate change to $u \in [u_0, U[$ where $u_0 = -a_0$ is small enough ($u_0 < U$). Then the 2 families given by $X^0, \overline{X}_1$ are similar (for ε or $\overline{\varepsilon} \to 0$ the singular point converges to a point outside the top of the parabola of singular points. It is the situation encountered in the blow-up at n_0 (on the spaces $P_{\overline{\lambda}_0}$ for $\overline{\lambda}_0 \neq (\pm 1, 0)$). The blow-ups :

$$
x = \omega x'' \quad y = \omega^2 y'' \quad \varepsilon = \omega^3 \varepsilon'' \qquad (51)
$$

along $\{a \in [-1/2, a_0]\}$ for X and :

$$
\overline{x} = \omega_1 x'' \quad \overline{y} = \omega_1^2 y_1'' \quad \overline{\varepsilon} = \omega_1^3 \varepsilon_1'' \qquad (52)
$$

along $\{u \in [0, U[\}$ for $\overline{X}_1$, desingularize X^2 along S_1.

The picture in each chart is similar to the one obtained at n_0 in the chart FR'1 (see figure 14) and the corresponding phase-directional chart.

Now the problem is to merge these 2 blow-ups in order to construct a single blown-up space M_3. The question of merging blow-ups performed in different charts, along a submanifold C and with a system of weights α, was studied in [DeR]. A sufficient condition is that the atlas of trivializations is an α-trivialization along C. Here $\alpha = (1, 2, 3)$, and the transition map given in (50) is linear and diagonal as a map from $(\overline{x}, \overline{y}, \overline{\varepsilon})$ to (x, y, ε); it trivially verifies the compatibility conditions of [DeR]. So, it is possible to define a new space M_3, with a blow-up map $\Phi_3 : M_3 \to M_2$, such that Φ_3 is given by the formulas (51), (52) in the respective local charts : $(x'', y'', \varepsilon'') \in S_+^2$, $\omega \in [0, \Omega[$ and $a \in [-1/2, a_0[$ for the first one and $(x_1'', y_1'', \varepsilon_1'') \in S_+^2$, $\omega_1 \in [0, \Omega[$ and

$u \in [0, U[$ for the second.

In the same way we blow-up S_2, S_1 and N_2. Let M_6, with projection $\pi_6 : M_6 \to \wedge$, be the resulting space and X^6 the resulting local field.

A typical section of M_6 (outside $\sum_1 \cup \sum_2$) is given in figure 19. This section is the 3-space Q_a, which is the closure in M_6 of $\pi_6^{-1}(m_a)$, where m_a is a line in parameter space corresponding to a fixed value of $a \neq 0$. (see also figure 5).

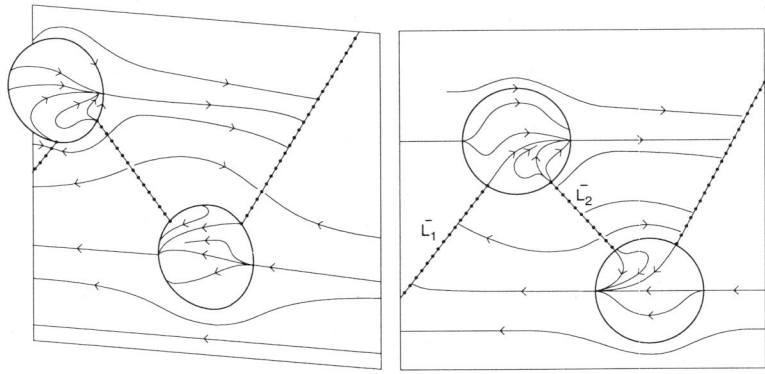

Figure 19
Phase portrait of $X^6|Q_a$.

2.5   Final blow-up

In M_6 all singular points are already elementary. But the collection of spaces $P_{\overline{\lambda}_0}$ is not a subfamily (disjointly) embedded in M_6 since each of them contains the singular fiber $\hat{F}_0$.

To arrange $\{P_{\overline{\lambda}_0}\}_{\overline{\lambda}_0}$ as a (disjoint) family, it suffices to blow-up along $\hat{F}_0$ in the direction of the parameter-space.

In fact X^6, which coincides with X^2 in a neighborhood of $\hat{F}_0$ is given by the three vector fields $\overline{X}_1$ in T_1, $\overline{X}_2$ in T_2 and X^0 in $M_0\backslash\{s_0, n_0\}$. Because the coordinate change between each of these charts are linear diagonal formulas of e.g. (ε, a) in function of $(\overline{\varepsilon}, \overline{a})$ in T_1 and (ε', a') in T_2, we can blow up with any weight. The best choice will be to take a weight 1 on a and 2 on ε such that the spaces $P_{\overline{\lambda}_0}$ will end up in the radial direction. For instance, in the chart T_1 we take :

$$\overline{a} = vA, \quad \overline{\varepsilon} = v^2 E \tag{53}$$

with $v \in [0, V[$ and $(A, E) \in S_+^1$.

If we replace this blow-up by the directional blow-ups with respectively $E = 1$, $A \in [-A_0, A_0]$ and $E \in [0, E_0[$, $A = \pm 1$, we obtain exactly the formulas (22), (23). As such $P_{\overline{\lambda}_0}$ corresponds to a fixed value of the parameter A or E, linked to $\overline{\lambda}_0$ by (21). The same is true in T_2, where formulas (41), (42) are now interpreted as blow-up formulas.

For the chart on $E_0 \backslash \{s_0, n_0\}$, we put

$$a = \tilde{v}\tilde{a}, \varepsilon = \tilde{v}^2 \tilde{\varepsilon} \tag{54}$$

with $\tilde{v} \in [0, \tilde{V}[$, $(\tilde{a}, \tilde{\varepsilon}) \in S_+^1$ in equations (5) of $X_{\varepsilon,a}$.

This chart may be replaced by directional ones with $\{\tilde{\varepsilon} = 1\}$ or $\{\tilde{a} = \pm 1\}$. Then, for instance, the change of coordinates between (22) and (54) with $\tilde{\varepsilon} = 1$ reduces in parameter direction to $\tilde{v} = uv$, $A = \tilde{a}$.

Finally we define our last blow up space M_7 with a blowing up map $\Phi_7 : M_7 \to M_6$. This map is given locally by maps like (53). Denote by $\tilde{F}_0 = \Phi_7^{-1}(\hat{F}_0)$ and $\tilde{\sigma}_i = \Phi_7^{-1}(\sigma_i)$ for $i = 1, 2$. Let X^7 be the resulting local vector field. In M_7, the family $\{P_{\overline{\lambda}_0}\}_{\overline{\lambda}_0}$ is embedded as a product, at least for $\overline{\lambda}_0 \neq (0, \pm 1)$. (For $\overline{\lambda}_0 = (0, \pm 1)$) we have performed an extra blow up along the lines (S_i, N_j). The local vector field X^7 provides a 1-parameter family of local vector fields $X^7 | P_{\overline{\lambda}_0}$. The equations for it are given in (28) near σ_1 and in (47) near σ_2 with the same parameter A (recall that $A = \dfrac{\overline{a}_0}{\sqrt{\overline{\varepsilon}_0}}$); elsewhere it is given by (54). We have drawn $X^7 | P_{\overline{\lambda}_0}$ for $\overline{\lambda}_0 = (1, 0)$ (i.e. $A = 0$) in the figure 17.

For the other values of A it suffices to replace the image on $D_{(1,0)}$ in figure 17 by the ones drawn in figure 7. (recall that $A = \overline{a}$). Our pictures are limited to $\partial M_7 = \pi^{-1}(\{\varepsilon = 0\})$. The next part is devoted to the study of $X^7 | P_{\overline{\lambda}_0}$ outside this critical locus.

In the sequel we will write M instead of M_7 and X instead of X^7, and denote the initial system by $X_{\varepsilon,a}$, where needed.

3 Foliations by center manifolds

The local vector field X as introduced in chapter 2 has non-isolated singular points. We denote the set of these points by $Z_d(X)$. As a consequence of the desingularization all these points are partially hyperbolic with one eigenvalue different from 0 and three eigenvalues equal to 0. It is easy to locate $Z_d(X)$ in the previous figures : $Z_d(X) \subset \partial M$. Later we will give a more precise description of these points.

From an adaptation, due to Bonckaert [Bon], of a well known result of Takens [T], we will deduce precise C^k normal forms for X near points in $Z_d(X)$. Next, these normal forms will be used to establish the existence of C^∞ foliations by center manifolds.

3.1 Normal forms for X at the non-isolated singular points

The normal form theorem that we will use at the partially hyperbolic singularities is the following :

Theorem 3 [Bon]
Let $X(y, x, z, a)$ on $\mathbb{R}^4$ be a C^∞ vector field having the following properties :

i) X is tangent to the foliation $da = 0$

ii) X is tangent to the foliation $dF(x, z) = 0$, where $F(x, z) = x^p z^q$ for $(p, q) = (0, 1)$ or $p, q \in \mathbb{N}_1$ and relatively prime.

iii) DX_0 has exactly one non-zero-eigenvalue and the related eigenspace is given by $\{x = z = a = 0\}$.

Let W be a C^k center manifold of X at 0, with $k \in \mathbb{N}_1$.

Then there exists a local C^k-coordinate change φ of the form $(y, x, z, a) \mapsto (\varphi_1(y, x, z, a), \varphi_2(y, x, z, a), \varphi_3(y, x, z, a), a)$ with

$$F(\varphi_2(y, x, z, a), \varphi_3(y, x, z, a)) = F(x, z),$$

and a strictly positive C^k function $f(y, x, z, a)$ such that

$$[f.\varphi_* X](y, x, z, a) = \pm y \frac{\partial}{\partial y} + Y(x, z, a) \qquad (55)$$

with Y of class C^k, $Y.a = 0$, $Y.F = 0$ and $\varphi(W) = \{y = 0\}$.

Remark : 1. We will say that the germ of X at p is C^k-equivalent to the right hand side of (55).

2. For the existence of C^k center manifold at p, we refer to [K] or to [HPS].

We now turn back to the points in $Z_d(X)$. To understand the structure of $Z_d(X)$ it is convenient to look at the figure 5, bearing in mind that the real dimensions are one unity bigger. $Z_d(X)$ is contained in ∂M, and it is an algebraic set contained in the closure of $\partial M \backslash (\Sigma_2 \cup \Sigma_1 \cup_i \hat{N}_i \cap_j \hat{S}_j)$; it contains strata of dimension 2, 1 and 0. Below we will only study the strata of dimension 2 and some of dimension 1.

The strata of dimension 2 are of 2 types :

$\boldsymbol{\alpha}$: the connected components of the union of the $int(L_i^-)$ for the lines L_i^- contained in $\hat{F}_{(0,a)}$, $a \neq 0$.

$\boldsymbol{\alpha'}$: The connected components of the union of the $int(L_i^-)$ for the lines L_i^- in $\tilde{F}_0$.

Each of these strata is diffeomorphic to $\mathbb{R}^2$ and fibered over the parameter a in case α, and over $\overline{\lambda}_0$ or A in case α', with as fiber the lines L_i^-.

The strata of dimension 1 that we need to study are :

$\boldsymbol{\beta}$: The connected components of the union of end point of the lines L_i^- of case α. They are contained in the interior of $\cup_i \partial \hat{N}_i \cup_j \partial \hat{S}_j$.

$\boldsymbol{\beta'}$: The connected components of the union of end points of the lines L_i^- of case β.

We can find local equations for X near each of these cases in the previous chapter. Let us recall them briefly :

Case $\boldsymbol{\alpha}$: X is represented by the initial family $X_{\varepsilon,a}$ given in (5) near a value $(0, a_0)$ with $a_0 \neq 0$, and a point $(x_0, y_0) \in \cup_i(int L_i)$. This means that $x_0 \neq 0, -1$ and $y_0 = \frac{x_0^2}{2} + \frac{x_0^3}{3}$, being the equation of L.

Case α' : X is represented by the 2-parameter family given in (54). More explicitly, we can take $\tilde{\varepsilon} = 1$ and perform the change $a = \tilde{v}\tilde{a}$, $\varepsilon = \tilde{v}^2$ to (5).
This gives the 2-parameter family :

$$\tilde{X}_{\tilde{v},\tilde{a}} : \begin{cases} \dot{x} = y - \dfrac{x^2}{2} - \dfrac{x^3}{3} \\[2mm] \dot{y} = \tilde{v}^2(\tilde{v}\tilde{a} - x) \end{cases} \tag{56}$$

We have to choose $\tilde{a}$ near some $\tilde{a}_0$, $\tilde{v}$ near 0 and (x,y) near $(x_0,y_0) \in \cup_i(int(L_i^-))$, which means $x_0 \neq 0, -1$ and $y_0 = \dfrac{x_0^2}{2} + \dfrac{x_0^3}{3}$.

Case β : Local equations are produced by blow-ups like (51). As in 2.4 we just treat the case of $S_1(\subset S)$. The points we want to consider are the end points of L_2^-, L_3^- obtained by blowing up S_1. We can study them in the directional blow up (51) with $y'' = 1$. The variables are $(x'', \omega, \varepsilon'')$ and a is a parameter. The blow up gives the one-parameter family :

$$X_a'' : \begin{cases} \dot{x}'' = 1 - \dfrac{x''^2}{2} - \dfrac{\omega}{3}x''^3 - \dfrac{1}{2}\varepsilon''x''(a - \omega x'') \\[2mm] \dot{\omega} = \dfrac{1}{2}(a - \omega x'')\varepsilon''\omega \\[2mm] \dot{\varepsilon}'' = -\dfrac{3}{2}(a - \omega x'')\varepsilon''^2 \end{cases} \tag{57}$$

We have a similar equation in the second chart used along S_1.

Case β' : It is the local vector field X in the family of spaces $P_{\overline{\lambda}_0}$, near the points s_2, s_3 and n_2, n_3.
They are represented by the one parameter family (parameter A) given in (28) for s_2, s_3 (chart PR1) and in (47) for n_2, n_3 (chart PR'1).
Now we will use theorem 3 above to obtain a simple normal form for X in each of the cases introduced above.

Proposition 4 (cases α, α').
In case α the germ of X is C^k equivalent, for any k, to $X_\varepsilon^N = \pm y\dfrac{\partial}{\partial y} + \varepsilon\dfrac{\partial}{\partial x}$. The local

diffeomorphism giving the equivalence sends the given point $(x_0, y_0, 0, a_0)$ to $(0, 0, 0, a_0)$ and preserves the (ε, a)-planes.

In case $\boldsymbol{\alpha'}$ the germ of X is C^k-equivalent, for any k to $X_{\tilde{v}}^N = \pm y \dfrac{\partial}{\partial y} + \tilde{v}^2 \dfrac{\partial}{\partial x}$. The local diffeomorphism giving the equivalence sends the given point $(x_0, y_0, 0, \tilde{a}_0)$ to $(0, 0, 0, \tilde{a}_0)$ and preserves $(\tilde{v}, \tilde{a})$-planes.

Proof : Take any point $(x_0, y_0, 0, a_0)$ of the type $\boldsymbol{\alpha}$. The vector field $X_{(0, a_0)}$ is normally hyperbolic along the line L. So, up to a translation and a linear change of coordinates we may assume to have coordinates (x, y) in which $(x_0, y_0) = (0, 0)$ and $j^1 X_{(0, a_0)}(0, 0) = \pm y \dfrac{\partial}{\partial y}$.

Moreover we have that $\dfrac{1}{\varepsilon}(X_{\varepsilon, a} - X_{0, a_0}) = Z_{\varepsilon, a}$ and Z_{0, a_0} is linearly independent of $\dfrac{\partial}{\partial y}$ at $(0, 0)$.

Now apply theorem 3. Given any k, $X_{\varepsilon, a}$ is C^{k+1}-equivalent to $X_1^N = \pm y \dfrac{\partial}{\partial y} + g(x, \varepsilon, a) \dfrac{\partial}{\partial x}$ where g is a C^{k+1} function.

The field X_1^N just like X must have an isolated 2-dimensional surface of singular points and up to an extra diffeomorphism in (x, ε, a) we may suppose that this surface is given by $\{\varepsilon = 0, y = 0\}$.

This implies that $g(x, \varepsilon) = \varepsilon \overline{g}(x, \varepsilon, a)$ with $\overline{g}(0, 0, 0) \neq 0$. (because of the expression in case $\boldsymbol{\alpha}$) Changing, if necessary, (x, y) into $(-x, -y)$, we can suppose that $\overline{g}(0, 0, 0) > 0$. The function $\overline{g}$ is C^k.

By an C^k-change of coordinates, respecting the (ε, a)-lines, we can change the family $\overline{g}(x, \varepsilon, a) \dfrac{\partial}{\partial x}$ into the constant one $\dfrac{\partial}{\partial x}$, yielding the result.

The case $\boldsymbol{\alpha'}$ is similar if we change ε by $\tilde{v}^2$ and a by $\tilde{a}$. $\qquad\square$

Proposition 5 (cases β, β')
In case β' the germ of X at the points s_i, $i = 2, 3$ and for $A = A_0$ is C^k equivalent, for any k, to the family :

$$X_A = \pm \left(z \frac{\partial}{\partial z} + v^2 f(u, v, A)(u \frac{\partial}{\partial u} - v \frac{\partial}{\partial v}) \right) \tag{58}$$

in some neighborhood of $(u, v, z, A) = (0, 0, 0, A_0)$.
The function f is C^k and strictly positive. The sign is positive for s_2 and negative

for s_3. Moreover $s_i = (0, 0, 0)$ and for each A the planes $\{v = 0\}$ and $\{u = 0\}$ are contained respectively in $\hat{F}_0$ and in the disk $D_{\overline{\lambda}}$ (recall that if $\overline{\lambda} = (\overline{\varepsilon}, \overline{a})$, $A = \dfrac{\overline{a}^2}{\overline{\varepsilon}^2}$).
The germ of X at the points n_i, $i = 1, 2$ and for $A = A_0$ is C^k equivalent for any k to the family

$$X_A = \pm \left(z' \frac{\partial}{\partial z'} + v'^2 f'(u', v', A)(u' \frac{\partial}{\partial u'} - 3v' \frac{\partial}{\partial v'}) \right) \tag{59}$$

with similar properties as above.
In case $\boldsymbol{\beta}$, we obtain a family on the parameter a, of a similar form as in (59).

Proof :
Consider the case β' for s_2, with the equations (28). We introduce the variable $z = +\sqrt{2} + \overline{x}$ such that s_2 is now located at $\{z = u = v = 0\}$ and the 1-jet of X (represented by the vector field family $\overline{X}_1$) is equal to $(\sqrt{2}z + \dfrac{2\sqrt{2}}{3}u)\dfrac{\partial}{\partial z}$. We can apply theorem 3 to find a C^{k+3} change of coordinates and parameters, tangent to the identity, which, up to multiplication by a C^{k+3} positive function changes $\overline{X}_1$ near $(z, u, v, A) = (0, 0, 0, A_0)$ into a family :

$$X_A^1 = z \frac{\partial}{\partial z} + g_1(u, v, A) \frac{\partial}{\partial u} + g_2(u, v, A) \frac{\partial}{\partial v} \tag{60}$$

with g_1 and g_2 of class C^{k+3}.
We can also assume that $\{u = 0\}$ and $\{v = 0\}$ are contained in the invariant spaces $D_{\overline{\lambda}_0}$ and $\hat{F}_0$ respectively and that expression (60) has the function uv as a first integral. Hence for $Z_A = g_1 \partial/\partial u + g_2 \partial/\partial v$ we get :

$$Z_A = g_3(u, v, A)(u \frac{\partial}{\partial u} - v \frac{\partial}{\partial v})$$

for a C^{k+2} function g_3.
This is clear since (from the invariance of $\{u = 0\}$ we can define g_3 by $g_1(u, v, A) = ug_3(u, v, A)$ and then observe, using $Z_A \cdot (uv) = 0$, that necessarily $g_2(u, v, A) = v \cdot g_3(u, v, A)$.
For each A the plane $\{z = 0\}$ is tangent to X_A^1 and $X_A^1|\{z = 0\} = Z_A$. For $\overline{X}_1$, this plane corresponds to an invariant submanifold W_A. The equations of (28) show that we can divide $\overline{X}_1|W_A$ by v^2 and obtain a hyperbolic vector field. But the plane $\{v = 0\}$, which corresponds to $\hat{F}_0$ and is the strong stable manifold of $\overline{X}_1$ and X_A^1,

is fixed by any change of coordinates, tangent to the idendity and sending $\overline{X}_1$ to X_A^1 (because it is easy to verify, using (28), that $\hat{F}_0$ is the unique C^1 invariant submanifold for $\overline{X}_1$, tangent to $\{v = 0\}$ at the point s_2. It follows that Z_A can be divided by v^2 to give a hyperbolic vector field : $g_3(u, v, A) = v^2 f(u, v, A)$ where f is a C^k positive function. This finishes the proof for s_2.

For the other points and cases the proof is completely similar, replacing the first integral uv by the first integral $u'^3 v'$ for the points n_i and by $\omega^3 \varepsilon''$ in the case $\boldsymbol{\beta}$. $\qquad\square$

3.2 Construction of center manifolds

In this part, we are going to construct center manifolds, one corresponding to each limit periodic set of "canard" type. Next we will gather these center manifolds in foliations, but let us first concentrate on the construction of a single one.

The existence of local center manifolds of class C^k at each $p \in Z_d(X)$ follows from general theorems (see [K, HPS, V] for instance). Using the propositions 4 and 5 above we can give a more precise construction.

We are particularly interested in center manifolds in a region $K = \cup_{\overline{\lambda}_0} \{P_{\overline{\lambda}_0} | - A_0 \leq A \leq A_0$ with $A = \dfrac{\overline{a}_0^2}{\overline{\varepsilon}_0}\}$ for A_0 small enough.

In this region, X is represented by a 1-parameter (say A- or $\overline{\lambda}_0$-) family of 3-dimensional local vector fields $X_A = X_{\overline{\lambda}_0} = X | P_{\overline{\lambda}_0}$ with local equations (24) or (28) in T_1, (43) or (47) in T_2, and (54) in chart R, with parameter $(x, y, \tilde{a}, \tilde{v})$. The non-isolated singular points of type $\boldsymbol{\alpha}'$ or $\boldsymbol{\beta}'$ have C^k normal forms as described in the propositions 4 and 5.

We will construct global center manifolds of dimension 3. Because the $P_{\overline{\lambda}_0}$ are invariant and transverse to the center direction, each center manifold $\mathcal{C}$ will be transverse to $P_{\overline{\lambda}_0}$ and hence induce in it a 2-dimensional center manifold $\mathcal{C}_{\overline{\lambda}_0} = \mathcal{C} \cap P_{\overline{\lambda}_0}$ for $X_A = X_{\overline{\lambda}_0} = X | P_{\overline{\lambda}_0}$. A C^∞ center manifold $\mathcal{C}$ is then the same as a C^∞ family of C^∞ center manifolds $\mathcal{C}_{\overline{\lambda}_0}$.

As, for each $\overline{\lambda}_0$, $P_{\overline{\lambda}_0}$ is a 3-dimensional space it will be possible to draw $\mathcal{C}_{\overline{\lambda}_0}$ in it. To make the presentation simpler we will consider $P_{\overline{\lambda}_0}$ as an embedded image in M of a unique 3-dimensional space P. This space is endowed with different charts (family rescaling chart with coordinates $(\overline{x}, \overline{y}, u)$, and equations (10), phase-directional rescaling chart with coordinates (v, θ, u) and equations (24) or $(v, \overline{x}, u)$ and equations (28), and so on) and a parameter respectively called $\overline{\lambda}_0$, A or $\tilde{a}$ $(\tilde{a} = A = \frac{\overline{a}_0}{\sqrt{\overline{\varepsilon}_0}})$.

$\partial P = \hat{F}_0 \cup \overline{D} \cup \overline{D}'$ where D, D' correspond to the disks $\overline{D}_{\overline{\lambda}_0}$, $\overline{D}'_{\overline{\lambda}_0}$. Of course we write σ_1 for $\partial \overline{D}$ and σ_2 for $\partial \overline{D}'$, as above, L_i^- for the blow up of the singular lines L_i and n_i, s_j for their end points (see chapter 2 for the exact description).

We choose a small rectangle T in P (see figure 20), with one side t in $D \subset \partial P$ and such that T is transverse to ∂P; T is chosen such that the orbit $\Gamma \subset D$ connecting s_2 and s_3, for $A = 0 = \overline{\lambda}_0$, is transverse to t at its middle point. Now let S be a sub-rectangle in T with a side s in $D \subset \partial P$ and $\Gamma \cap t$ being the middle of s. We suppose that $A_0(> 0)$ as well as T are chosen small enough, such that X_A is transverse to T for all A with $-A_0 \le A \le A_0$ and at any point of T, and such that the separatrix $\Gamma_s(A)$ of s_2 and the separatrix $\Gamma_u(A)$ of s_2, for the local vector field X_A, cut T in the interior of s. We want to prove that, if S and A_0 are small enough, a Poincaré map h_A for X_A, with $-A_0 \le A \le A_0$, is defined from S to T. In fact the construction of the center manifold will be used to prove the existence of such a map and to study its recurrence properties, i.e. to study the limit cycles that cut S. Those will be the limit cycles for (a, ε) near $(0, 0)$.

3.2.1   Center manifolds of type I

Let $\gamma(\tilde{v}) = (\tilde{v}, x(\tilde{v}), y(\tilde{v}))$ be a C^∞ arc in the chart R (recall that this chart is given by $(x, y) \in I\!\!R^2 \backslash \{n_0, s_0\}$ and the parameters $\tilde{v}, \tilde{a}$ with $\varepsilon = \tilde{v}^2$ and $a = \tilde{a}\tilde{v}$. Moreover $\tilde{a} = A$, so that $(x, y, \tilde{v})$ are the coordinates in P). We suppose that $\tilde{v} \in [0, \tilde{v}_0[$ and that $(x(0), y(0)) \in C_1$ where C_1 is the region : $C_1 = \{\frac{x^2}{2} + \frac{x^3}{3} < y < \frac{1}{6}, x > -1\}$. Let us denote by $\mathcal{C}_\gamma(A)$ the saturate of γ, more specifically the closure of the union of segments of orbits of X_A through the points of graph(γ), and taken between the first intersection of this trajectory with S in negative time and with T in positive time. (As far as such points exist).

Theorem 6 : *If A_0 and $\tilde{v}_0$ are small enough, then $\mathcal{C}_\gamma(A)$ is well defined for $-A_0 \le A \le A_0$. It is a C^∞ manifold (with boundary) outside the 2 points $\alpha(\gamma) \in L_1^-$, $\beta(\gamma) \in L_2^-$, limit points of the trajectory of $X_{(0,0)}$ through $(x(0), y(0))$; $\mathcal{C}_\gamma(A) \backslash \{\alpha(\gamma), \beta(\gamma)\}$ depends C^∞-ly on A.*

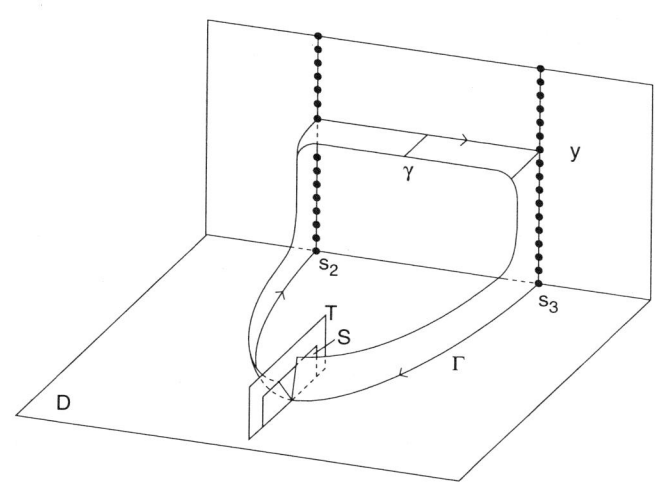

Figure 20
A typical center manifold $\mathcal{C}_\gamma(A)$.

Remark : Near each point in the segments $]\alpha(\gamma), s_2] \subset L_1^-$ and $]\beta(\gamma), s_3] \subset L_2^-$, $\mathcal{C}_\gamma = \cup_A \mathcal{C}_\gamma(A)$ is a C^∞ center manifold.

Proof : Of course, along each orbit contained in $\mathcal{C}_\gamma(A)$ and passing through graph(γ), the manifold $\mathcal{C}_\gamma(A)$ is C^∞. We just have to prove that $\mathcal{C}_\gamma(A)$ is also C^∞ along the segments $]\alpha(\gamma), s_2] \subset L_2^-$, $]\beta(\gamma), s_3] \subset L_3^-$ and the segments of separatrices $\Gamma_2(A), \Gamma_3(A)$ of s_2, s_3 until their intersection with t (recall that $\Gamma_2(0) \cup \Gamma_3(0) = \Gamma$).
We will prove that $\mathcal{C}_\gamma(A)$ is C^k for any k.
Let us start with the segment $]\alpha(\gamma), s_2[$. Take a cover of $[\alpha(\gamma), s_2]$ by open sets in which X_A has the C^k normal form given by the propositions 4 or 5. By the flow of $-X_A$, we can push a small segment $\gamma[0, \tilde{v}_1]$, of the graph of γ, for some $0 < \tilde{v}_1 < \tilde{v}_0$, inside the first open set U_1 (the one covering $\alpha(\gamma)$). This image γ_1 of $\gamma([0, \tilde{v}_1])$ is a C^∞ arc in the initial coordinates and a C^k arc in the normal coordinates in U_1. The C^k-smoothness of $\mathcal{C}_\gamma(A)$ along $]\alpha(\gamma), s_2[$ easily follows from the following lemma.

Lemma 7

Take a C^k system of coordinates $(x, y, \tilde{v}, A)$ on an open set U in which X_A has the expression $X_A = y\dfrac{\partial}{\partial y} + \tilde{v}^2\dfrac{\partial}{\partial x}$. Take any C^k family of arcs $\tilde{\gamma}(\tilde{v}, A) = (\tilde{v}, x(\tilde{v}, A), y(\tilde{v}, A))$ with $(\tilde{v}, A) \in [0, \tilde{v}_1] \times [-A_0, A_0]$, $\tilde{v}_1 > 0$ small enough (such that the image of $\tilde{\gamma}$ is inside U). Then the set $\mathcal{C}_{\tilde{\gamma}}(A)$ obtained as closure of the saturation of $\mathrm{graph}(\tilde{\gamma})$ by the orbits of $-X_A$, is C^k along the set $\{y = 0, x < x(0, A)\}$, and so is C^k everywhere outside the points $\{x(0, A), 0\}$.

Proof of lemma 7

The graph of $\mathcal{C}_{\tilde{\gamma}}(A)$ is explicitely given by :

$$y(\tilde{v}, x, A) = y(\tilde{v}, A)e^{\frac{x - x(\tilde{v}, A)}{\tilde{v}^2}} \tag{61}$$

which is a C^k function for $x < x(0, A)$. □

At this point we have proved that $\mathcal{C}_{\gamma}(A)$ is C^k all along $]\alpha(\gamma), s_2[$. Take then normal form coordinates on U around s_2, in which X_A takes the form :

$$X_A = z\frac{\partial}{\partial z} + v^2 f(u, v, A)[u\frac{\partial}{\partial u} - v\frac{\partial}{\partial v}] \tag{62}$$

with $U = [-z_0, z_0] \times [u_0, u_0] \times [-v_0, v_0] \times [-A_0, A_0]$.

The space P corresponds to the quadrant $\{u \geq 0, v \geq 0\}$. Take u_1, $0 < u_1 < u_0$.

The surface $\mathcal{C}_{\gamma}(A)$ cuts $\{u = u_1\}$ along a curve $\tilde{\gamma}$, graph of a C^k function $z(v, A)$. (In fact it follows from (61) that all the derivatives of z are flat at $v = 0$, but we do not use it here in order to prove the following Lemma, which implies the C^k-smoothness of $\mathcal{C}_{\gamma}(A)$ in a neighborhood of $s_2 = (0, 0, 0) \in P$.

Lemma 8 : *Let X_A be given by (62) and take a C^k function $z(u, A)$ representing a curve $\tilde{\gamma}$ in $\{u = u_1\}$. The set $\mathcal{C}_{\tilde{\gamma}}$ obtained as closure of the saturation of the graph $\tilde{\gamma}$ by the orbits of $-X_A$, is C^k along the axis $\{u = z = 0\}$ and so is C^k everywhere.*

Proof of the lemma 8

We are looking at the graph $Z(u, v, A)$ of $\mathcal{C}_{\tilde{\gamma}}$ (for $u > 0$, $v > 0$. We extend Z by

$Z(0, v) = Z(u, 0) = 0)$.

To obtain the result we integrate the equation of $-\dfrac{1}{v^2 f} X_A$:

$$\begin{cases} \dot{z} = -\dfrac{1}{v^2 f} z = -\dfrac{F}{v^2} z \\[4mm] \dot{u} = -u \\[4mm] \dot{v} = v \end{cases} \qquad (63)$$

with initial condition on $\tilde{\gamma}$: $(z(v_0, A), u_1, v_0)$. (We suppose that $f(u, v, A) > 0$ on the domain U and write $\dfrac{1}{f} = F$). The integration of (63) gives :

$$\begin{cases} v_t = v_0 e^t \\ u_t = u_1 e^{-t} \\ z_t = z(v_0, A) exp\left[-\int\limits_0^t \dfrac{F(u_\tau, v_\tau, A)}{v_\tau^2} d\tau \right] \end{cases} \qquad (64)$$

We now fix a value (u, v) with $0 < u < u_1$ and $v > 0$. We have

$$\begin{cases} v = v_0 e^t \\ u = u_1 e^{-t} \end{cases} \qquad (65)$$

Hence $t = ln\dfrac{u_1}{u}$ and $v_0 = \dfrac{uv}{u_1}$. The function $Z(u, v, A)$ is then given by :

$$Z(u, v, A) = z(\dfrac{uv}{u_1}, A) exp\left[-\left(\dfrac{u_1}{uv}\right)^2 \int_0^{ln\frac{u_1}{u}} e^{-2\tau} F(u_1 e^{-\tau}, \dfrac{uv}{u_1} e^\tau, A) d\tau \right] \qquad (66)$$

We make the change of variable $s = e^{-\tau}$ in the integral term of (66) :

$$\int_0^{ln\frac{u_1}{u}} \ldots d\tau = \int_{u/u_1}^1 sF(u_1 s, \dfrac{uv}{u_1 s}, A) ds$$

And finally, the explicit expression of Z :

$$Z(u, v, A) = z(\dfrac{uv}{u_1}, A) exp\left[-\left(\dfrac{u_1}{uv}\right)^2 \int_{u/u_1}^1 sF(u_1 s, \dfrac{uv}{u_1 s}, A) ds \right] \qquad (67)$$

It is easy to verify that Z is C^k for $v \geq 0$, $u_1 > u \geq 0$, once extended by 0 on $\{uv = 0\}$.

It suffices to verify that all partial derivatives of order at most k have a continuous extension by 0 on $\{uv = 0\}$.

First, if $F_0 = \inf\{|F(u, v, A)| | (u, v, A) \in U\} > 0$ we have for $0 < u < \overline{u}_1 < u_1$ the existence of a constant $c > 0$ depending on u_1 with

$$|Z(u, v, A)| \leq |z(\frac{uv}{u_1}, A)| \cdot exp(-cF_0 \left(\frac{u_1}{uv}\right)^2) \tag{68}$$

Next observe that $Z(u, v, A) = \overline{Z}(u, w, A)$ with :

$$\overline{Z}(u, w, A) = z(w, A)exp\left[-\frac{1}{w^2} \int_{u/u_1}^1 sF(u, s, \frac{w}{s}, A)ds\right] \tag{69}$$

and $w = uv/u_1$. It suffices to prove the property that each derivative of $\overline{Z}$, up to order k tends to zero for $u \to 0$.

We can suppose that $u \geq w$ (i.e. $v \leq 1$). It is easy to see that :

$$\frac{\partial^{l_1+l_2+l_3}}{\partial w^{l_1} \partial u^{l_2} \partial A^{l_3}} \left[\int_{u/u_1}^1 sFds\right] \leq \frac{*}{w^{l_1-1}}, \tag{70}$$

where $*$, here and in the sequel, symbolizes a constant depending just on the k-norm of F on U. Note by $|Z|$ the l-norm of z. An easy induction using (70) on the form of the derivative leads to the estimates implying the required property.   □

We have proved that $C_\gamma(A)$ is C^k at s_2 and so along an interval of the separatrix $\Gamma_2(A)$ of s_2, arriving at s_2 (it is represented by the v-axis in the normal coordinates, that we have used in lemma 7). The C^k-smoothness of $C_\gamma(A)$ for the remaining part of $\Gamma_2(A)$ is now trivial because the rest of $C_\gamma(A)$ is obtained by saturation by the regular C^∞ flow of $-X_A$ near $\Gamma_2(A)$.

Of course, the proof of the C^k-smoothness of the part of $C_\gamma(A)$ obtained by saturation by the flow of X_A (for positive t) is completely similar to the one above. Finally $C_\gamma(A)$ was shown to be C^k (outside $\alpha(\gamma)$, $\beta(\gamma)$) for any k, and hence is C^∞.   □

3.2.2   Center manifolds of type II

Let now γ be a C^∞ arc in P, ending at D'. More precisely, we suppose that the end point m of γ for $A = 0$ is contained in D'_0 belonging to the basin of expansion C_3 of the source n_3 : in the figure 14 it is the simply connected open domain of the disk D', whose boundary is the union of the arc $n_2 n_3 n_4$ on $\partial D'$ and of the separatrix of n_2, connecting n_2 and n_4. We limit $A \in [-A_0, A_0]$ in a way that m belongs to the basin of expansion of n_3 for any A. Now, as above we call $C_\gamma(A)$, the closure of the union of the segments of trajectories of X_A, through the points of graph(γ), and between S and T.

Theorem 9 : *If γ and A_0 are small enough, $C_\gamma(A)$ is well defined for $-A_0 \leq A \leq A_0$. It is a C^∞ manifold outside the horizontal segment $[n_4, n_4']$, closure of an orbit in $\hat{F}_0$ which is contained in $C_\gamma(A)$. Outside these points, $C_\gamma(A)$ depends C^∞-ly on a.*

Remark : $C_\gamma = \cup_A C_\gamma(A)$ is a C^∞ center manifold at each point of $Z_d(X)$ belonging to it, except for n_4 and n_4'.

Proof.

Starting at γ we defined $C_\gamma(A)$ by following the flow of X_A through γ, in positive and negative time. In particular a part of the boundary of $C_\gamma(A)$ is the orbit of the end point m of γ in D'. $C_\gamma(A)$ is certainly C^∞ along this orbit and the first difficulties show up at the limit points n_3 and n_4 (see figures 14, 17).

i) Passing the point n_3

We will prove that $C_\gamma(A)$ is C^k for any k at n_3 and so, is C^∞. Therefore, take any k and choose C^k normal form coordinates around n_3 for X_A as given in proposition 5, formula (59). If we replace z', u', v' by z, u, v we have

$$X_A = z\frac{\partial}{\partial z} - v^2 f(u, v, A)(u\frac{\partial}{\partial u} - 3v\frac{\partial}{\partial v}).$$

We reduce the chart U to $U = [-z_0, z_0] \times [-v_0, v_0] \times [-u_0, u_0] \times [-A_0, A_0]$ where $f(u, v, A) > 0$ everywhere.

Now we can push γ inside U, taking, if necessary, a small arc of γ at m and cutting $C_\gamma(A)$ in U by a transverse plane $\{v = v_1\}$ with $v_0 > v_1 > 0$. In the plane $\{v = v_1\}$ we obtain a C^k arc $z(\tilde{u}, A)$ with $\tilde{u} \in [0, u_1]$ and $u_1 < u_0$ contained in $C_\gamma(A)$. The C^k smoothness of $C_\gamma(A)$ at n_3 follows from :

Lemma 10 : *Let a C^k function $z(\tilde{u}, A)$ define an arc $\tilde{\gamma}$ in the plane $\{v = v_1\}$ of the coordinate chart U defined above. The set $C_{\tilde{\gamma}}$ obtained as a closure of the saturation of $\tilde{\gamma}$ by the orbits of $-X_A$ is C^k along the axis $\{v = z = 0\}$ and so is C^k*

everywhere.

Proof of the lemma 10

We consider the equations of $-\dfrac{X_A}{fv^2}$ for $v > 0$:

$$
\begin{cases}
\dot{z} = -\dfrac{F}{v^2} z \\[2mm]
\dot{u} = u \\[2mm]
\dot{v} = -3v
\end{cases}
\tag{71}
$$

with $F = \dfrac{1}{f}$. Let $F_0 = \inf_U F(u,v,A) > 0$.

The integration of (71) gives :

$$
u_t = \tilde{u}e^t \quad v_t = v_1 e^{-3t} \quad \text{and} \quad z_t = z(\tilde{u}, A)exp[-\int_0^t \frac{F}{v_\tau^2} d\tau]
\tag{72}
$$

Consider a value (u,v) with $u > 0$, $v_1 > v > 0$.

We can compute : $3t = ln\dfrac{v_1}{v}$, $u^3 v = \tilde{u}^3 v_1$ and if we eliminate $\tilde{u}, t$ in (72) for v, u we obtain a function $Z(u,v,A)$ whose graph is precisely $\mathcal{C}_\gamma(A)$ (for $u > 0$, $v_1 > v > 0$) :

$$
Z(u,v,A) = z\left(\frac{uv^{1/3}}{v_1^{1/3}}, A\right) exp\left[-\frac{1}{v_1^2}\int_0^{1/3\ln\ v_1/v} e^{6\tau} F\left(\frac{uv^{1/3}}{v_1^{1/3}}e^\tau, v_1 e^{-3\tau}, A\right) d\tau\right]
\tag{73}
$$

Again, putting $s = e^{3\tau}$ in the integral and introducing $\omega = \dfrac{uv^{1/3}}{v_1^{1/3}}$, we find that

$Z(u,v,A) = \overline{Z}(\omega, v, A)$ given by :

$$
\overline{Z}(\omega, v, A) = z(\omega, A)exp\left[-\frac{1}{3v_1^2}\int_1^{v_1/v} sF(\omega s^{1/3}, \frac{v_1}{s}, A)ds\right]
\tag{74}
$$

It suffices to prove the flatness of each derivative of $\overline{Z}$ for $\{v = 0\}$.

As in the lemma 8 we can first establish by an induction on (l_1, l_2, l_3) that for all (l_1, l_2, l_3) with $l = l_1 + l_2 + l_3 \leq k$ we have

$$
\left|\frac{\partial^l}{\partial^{l_1}\omega \partial^{l_2} v \partial^{l_3} A}\left[\int_1^{v_1/v} sF(\omega s^3, v_1/s, A)ds\right]\right| \leq \frac{*}{v^{3l_1+2}},
\tag{75}
$$

where $*$ is a constant which is related to the l-norm of F. From this easily follow the required properties. $\square$

ii) Passing the point n_4

The point is hyperbolic for each A, with a fixed spectrum of eigenvalues. Equations are given by (43) in coordinates (θ', v', u').

The eigenvalues are $-1, -3/4$ (in the direction of D' with coordinates θ', v') and $+1/4$ (in the direction of $\hat{F}_0$ with coordinate u').

The problem is the following : we have some C^k arc $\tilde{\gamma}$ transverse to D' at its end point $\tilde{m} \in C_3$. We want to saturate it for positive times to obtain the continuation of $C_\gamma(A)$ along the unstable separatrix $n_4 n_4'$ of n_4. In general, due to the resonance between the eigenvalues at n_4 there is no hope to obtain a C^k smoothness of (the closure of) this extension along the unstable separatrix.

We however only need the following estimates.

Lemma 11 : *Let a C^k arc $\tilde{\gamma}$ as above be defined in a coordinate system U' around n_2. Then the saturated set $C_{\tilde{\gamma}}(A)$ of $\tilde{\gamma}$ in positive time is a C^k manifold outside $[n_4, n_4']$. For any $u_1' > 0$, its intersection with a plane $\{u' = u_1\}$, in the domain of U', is the graph of a curve $\theta'(v', A)$ which for some $r \in \mathbb{R}$ verifies following estimates : for all $l_1, l_2, l = l_1 + l_2 \leq k$:*

$$\frac{\partial^l}{\partial v'^{l_1} \partial A^{l_2}} \theta'(v', A) = 0(v'^{r-l}) \tag{76}$$

Proof : It is not so hard to obtain the required estimates by using the variational equation. Taking into account that the eigenvalues in the u', θ' and v'-direction are respectively $1/4, -1$, and $-3/4$ (see (44)) one can take $r = 4/3$. $\qquad\square$

Remark : The conditions (76) are invariant by any diffeomorphism in the (v', θ') plane preserving $(0,0)$ and the θ'-axis. So they do not depend on the choice of coordinates.

iii) Passing n_4'

All along $[n_4, n_4'[$, $C_\gamma(A)$ has the kind of smoothness expressed in (76) : taking any coordinates $(x, \tilde{v})$, transverse to $]n_4, n_4'[$ at a point $m \in]n_4, n_4'[$, with $m = (0,0)$ and $\{\tilde{v} = 0\}$ in $\hat{F}_0$, $C_\gamma(A)$ cuts this plane of coordinates along a curve $x(\tilde{v}, A)$ with estimates like in (76).

Now take a C^k system $(x, \tilde{v}, y)$ of normal coordinates at n_4' where $X_A = -y\frac{\partial}{\partial y} + \tilde{v}^2\frac{\partial}{\partial x}$.

Beyond the point n'_4 (the closure of) the continuation of $C_\gamma(A)$ will again be C^k along the set of zeroes $\{y = \tilde{v} = 0\}$.

The graph of $C_{\tilde{\gamma}}(A)$ is explicitely given by :

$$y(\tilde{v}, x, A) = y_0 e^{\frac{x - x(\tilde{v}, A)}{\tilde{v}^2}} \tag{77}$$

where $y_0 > 0$; we suppose that $\tilde{\gamma}$ is a curve $\{y = y_0, x = x(\tilde{v}, A)\}$.

The estimates

$$\frac{\partial^l x}{\partial^{l_1} \tilde{v} \partial^{l_2} A}(\tilde{v}, A) = 0(\tilde{v}^{r-l}) \tag{78}$$

which we have on $x(\tilde{v}, A)$, easily imply that $y(\tilde{v}, x, A)$ is C^k with flat derivatives of order $\leq k$ along the line L_3^- locally given by $\{y = \tilde{v} = 0\}$ and for $x > x(0, A)$, the last value being the coordinate of n'_4.

iv) Final touch to the proof of Theorem 9

The continuation of $C_\gamma(A)$ as a C^k manifold beyond n'_4 in positive time and beyond n_3 in negative time is completely similar to the one made for the center manifolds of type I and can be omitted.

Of course $C_\gamma(A)$, being C^k for any k, is C^∞ outside the line $[n_4, n'_4]$.   □

3.2.3   Center manifolds of type III

Let $\gamma(\tilde{v}) = (x(\tilde{v}), y(\tilde{v}))$ be a C^∞ arc in the chart R, defined for $\tilde{v} \in [0, \tilde{v}_0[$ and such that $m = (x(0), y(0)) \in C_2$ where C_2 is the region :

$$C_2 = \{(x, y) \in I\!R^2 | 0 < y < \frac{1}{2}x^2 + \frac{1}{3}x^3, x < 0\}. \tag{79}$$

Let $C_\gamma(A)$, as above, be the closure of the union of orbit-segments saturating the graph of γ, from S to T. We have :

Theorem 12 : *Suppose that A_0 and $\tilde{v}_0$ are small enough. Then $C_\gamma(A)$ is well defined. Its intersection with ∂P consists of the union of segment of orbits in ∂P and of the singular lines L_i^- (see figure 21). $C_\gamma(A)$ is C^∞ outside $\alpha(\gamma), \beta(\gamma)$ (limit points of the orbit of m in $\hat{F}_0$) and the segment $[n_4, n'_4]$, and depends C^∞-ly on A outside these points.*

Proof : The details of the proof are completely similar to those described above : smoothness along the lines L_i^- follows from lemma 7, at the point n_1 from a similar argument as in lemma 10, and the behaviour along $[n_4, n_4']$ follows from lemma 11.□

3.2.4 Pictures of the center manifolds

For $A = 0$ each type of center manifold cuts ∂P along a canard-like periodic limit set, respectively corresponding to the drawings (2), (3) and (4) in figure 3 for the center manifolds of type I, II and III.

In figure 21 we give a picture of the "blow down" of these center manifolds in the 3-dimensional space $\pi_0^{-1}(l_{(1,0)})$ (recall that $\overline{\lambda}_0 = (1,0)$ corresponds to $A = 0$).

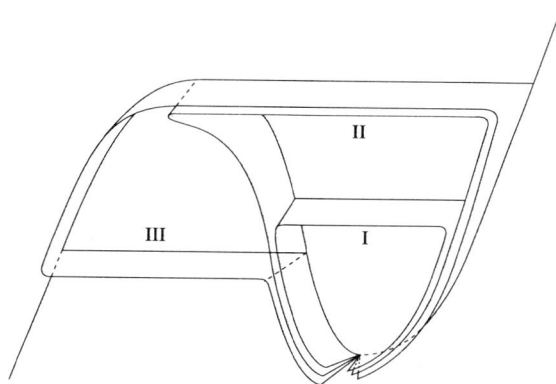

Figure 21

Center manifolds of type I, II, III in $\pi_0^{-1}(l_{(1,0)})$.

3.3   Foliations by center manifolds

Let us now look how the center manifolds - constructed above - can be gathered into foliations.

3.3.1   Foliation of type I

To construct the foliation we consider as initial condition a C^∞ family of segments γ, foliating a rectangle $\mathcal{G}_I \subset P$ with the property (see also figure 22) that $\mathcal{G}_I \cap \hat{F}_0 = [\alpha, \beta]$, $\mathcal{G}_I \cap \sigma_1 = \{\beta\}$, $\mathcal{G}_I \cap \sigma_2 = \{\alpha\}$ and $\mathcal{G}_I$ has sides on respectively D and D'. The foliation given by the γ inside $\mathcal{G}_I$ is transverse to $[\alpha, \beta]$ and ends in respectively D and D'.

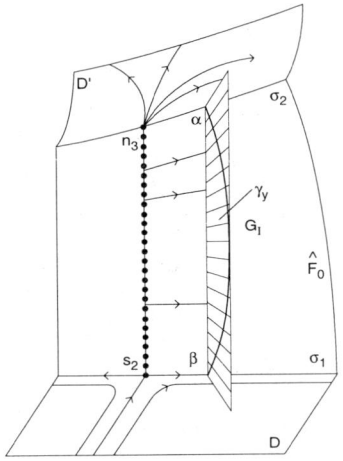

Figure 22

$\mathcal{G}_I \subset P$, *foliated by segments* γ_y.

By using the blow-up mapping $\pi : \hat{F}_0 \to I\!\!R^2$ we can parametrize $[n_3, s_2]$ as well as $[\alpha, \beta]$ by the variable $y \in [0, 1/6]$, with $y(s_2) = y(\beta) = 0$ and $y(n_3) = y(\alpha) = 1/6$. Let us write γ_y for the segment cutting $\hat{F}_0$ at a point with parameter y.

For each $y \in]0, 1/6[$ we consider the saturation $\mathcal{C}_{\gamma_y}(A)$ as defined in 3.2.1; we call it $\mathcal{C}_I(y, A)$. Like in 3.2.1 we take $A \in [-A_0, A_0]$ for some $A_0 > 0$ sufficiently small such that $\mathcal{C}_I(y, A)$ will cut T in positive time along a curve $\{h = F_I(u, y, A)\}$ and in negative time along a curve $\{h = B_I(u, y, A)\}$. For the moment (u, h) represent C^∞ variables on T with $T \cap D = \{u = 0\}$ and such that $\{(A, u, h) = (0, 0, 0)\}$ corresponds to the point on the connection Γ between s_2 and s_3.

More precisely u is one of the coordinates as used in (9), and later on we will give a more precise meaning to h; but in any case and with respect to the coordinates in (9), it is an increasing regular parameter for the reversed $\bar{y}$-axis.

We saw in 3.2.1, that for fixed y, the functions $F_I(u, y, A)$ and $B_I(u, y, A)$ are C^∞. Similar calculations permit to show that these functions are C^∞ in all variables, including y.

Of course, when y tends to 0 or $1/6$, the domain of definition in u tends to zero. The functions F_I and B_I are C^∞ and defined on a domain

$$\mathcal{D}_I = \{(u, y, A)| - A_0 \leq A \leq A_0, \ 0 \leq y \leq 1/6, \ 0 \leq u \leq u_I(y)\} \tag{80}$$

where $u_I(y)$ is a continuous function with $u_I(y) > 0$ for $0 < y < 1/6$ and $u_I(0) = u_I(1/6) = 0$.

We are now going to investigate flatness properties of F_I and B_I. We could do this among the $\mathcal{C}_I(y, A)$ only, but in view of the proof of theorem 1 we prefer to present the flatness with respect to a reference "center manifold" $\mathcal{C}_{II}(A)$ of type II, as constructed in 3.2.2. We denote its intersection with T in positive and negative time respectively by $\{h = F_{II}(u, A)\}$ and $\{h = B_{II}(u, A)\}$. The functions F_{II} and B_{II} are also C^∞.

Proposition 13

In terms of the notation introduced above there exist C^∞ functions $k_B(y, A)$ and $k_F(y, A)$ on $]0, 1/6[\times[-A_0, A_0]$ with $\dfrac{\partial k_B}{\partial y}(y, A) > 0$ and $\dfrac{\partial k_F}{\partial y}(y, A) > 0$ such that

$$B_{II}(u, A) - B_I(u, y, A) = \exp\left(-\frac{k_B(y, A)}{u^2}(1 + \varphi_B(u, y, A))\right) \tag{81}$$

$$F_{II}(u, A) - F_I(u, y, A) = \exp\left(-\frac{k_F(y, A)}{u^2}(1 + \varphi_F(u, y, A))\right) \tag{82}$$

where φ_B and φ_F are C^∞ on some $\mathcal{D}_I \cap \{u > 0\}$, and all partial derivatives of φ_B and φ_F with respect to (y, A) as well as $u\dfrac{\partial \varphi_B}{\partial u}(u, y, A)$ and $u\dfrac{\partial \varphi_F}{\partial u}(u, y, A)$ are $O(u)$ for $u \to 0$, uniformly in (y, A) on any compact subset of $]0, 1/6[\times [-A_0, A_0]$.

Before proving proposition 13 it is interesting to make a few remarks that we incorporate in the following lemma.

Lemma 14

Let $\ell \in \mathsf{N}_3 \cup \{\infty\}$. On the space of germs of functions $h(u)$, with $h(0) = 0$ let us consider the "graph-transform", defined by the action of the group of germs at $(0,0)$ of C^ℓ diffeomorphisms $G(u, h)$, respecting $\{u = 0\}$ in an orientation-preserving way. Let h be such that $G(\text{graph } h) = \text{graph } (\tilde{h})$. Take $M > 0$, $N > 0$ and $k(\lambda) > 0$. Then, if $h_1(u, \lambda)$ and $h_2(u, \lambda)$ are two C^ℓ λ-families of germs with

$$\begin{cases} |h_i(u, \lambda)| \leq Mu \\ h_2(u, \lambda) - h_1(u, \lambda) = \exp\left(-\dfrac{k(\lambda)}{u^2}(1 + \varphi(u, \lambda))\right) \\ \left|\dfrac{\partial^{|j|}\varphi}{\partial \lambda^j}(u, \lambda)\right| \leq Nu, \;\; for \; 0 \leq |j| \leq \ell, \; and \; \left|\dfrac{\partial \varphi}{\partial u}(u, \lambda)\right| \leq N \end{cases} \tag{83}$$

we necessarily have

$$\begin{cases} |\tilde{h}_i(u, \lambda)| \leq \tilde{M}u \\ \tilde{h}_2(u, \lambda) - \tilde{h}_1(u, \lambda) = \exp\left(-\dfrac{ck(\lambda)}{u^2}(1 + \tilde{\varphi}(u, \lambda))\right) \\ \left|\dfrac{\partial^{|j|}\tilde{\varphi}}{\partial \lambda^j}(u, \lambda)\right| \leq \tilde{N}u, \;\; for \; 0 \leq |j| \leq \ell, \; and \; \left|\dfrac{\partial \tilde{\varphi}}{\partial u}(u, \lambda)\right| \leq \tilde{N} \end{cases}$$

and this for some $c > 0$ only depending on G, while $\tilde{M} = \tilde{M}(M, G) > 0$ and $\tilde{N} = \tilde{N}(N, M, G) > 0$.

Proof. Let us first consider a diffeomorphism G of the form $G(u, h) = (u, G_2(u, h))$. Then $\tilde{h}_i(u) = G_2(u, h_i(u))$.

We can develop $G_2(u, h) = \varphi_0(u) + h\varphi_1(u) + \varphi_2(u, h)$ with $\varphi_1(0) > 0$, φ_1 and φ_2 of class C^ℓ with $\varphi_2(u, h) = O(h^2)$.

One can find some $\psi(u, v_1, v_2)$ of class $C^{\ell-2}$, such that

$$\varphi_2(u, h_1) - \varphi_2(u, h_2) = \psi(u, h_1, h_2)(h_2 - h_1)^2$$

By this

$$(\tilde{h}_2 - \tilde{h}_1)(u, \lambda) = (h_2 - h_1)(u, \lambda)[\varphi_1(u) + O(h_2 - h_1)]$$
$$= (h_2 - h_1)(u, \lambda)\exp\left[\,\ln(\varphi_1(u) + O(h_2 - h_1))\right]$$

yielding the result. In this special case $c = 1$, $\tilde{M}$ depends on M and φ_0, and $\tilde{N}$ can be any number strictly bigger than N.

In the rest of the calculation we can limit to $h_1 = 0$, and to a G respecting $\{u = 0\}$ and $\{h = 0\}$.

Such a G can be written as a succession of the two diffeomorphisms :

$$G^1(u, h) = (u, h \cdot G_2^1(u, h))$$
$$G^2(u, h) = (u \cdot G_1^2(u, h), h)$$

There only remains to consider the action of G^2, with $G_1^2(0, 0) > 0$.
Let $(w, h) \to (w \cdot H(w, h), h)$ be the inverse diffeomorphism of G^2. Then $\tilde{h}_2(w)$ is defined by the implicit equation :

$$h - \exp\left(-\frac{k(\lambda)}{w^2 \cdot (H(w, h))^2}(1 + \varphi(w \cdot H(w, h), \lambda))\right) = 0$$

that we can also write as

$$h - \exp\left(-\frac{ck(\lambda)}{w^2}(1 + \psi(w, h, \lambda))\right) = 0 \tag{84}$$

with $c = (H(0, 0))^{-2}$, ψ defined in an implicit way out of φ and of $(H(0, 0))^{-2} \cdot (H(w, h))^2$.

The derivative, at $(0, 0)$, of expression (84) being 1, we can use the implicit function theorem to get $h = \tilde{h}_2(w) = \tilde{h}(w)$.

If we substitute $\tilde{h}$ in (84) we get

$$\tilde{h}(w) = \exp\left(-\frac{ck(\lambda)}{w^2}(1 + \tilde{\varphi}(w, \lambda)\right)$$

with $\tilde{\varphi}$ having the required properties with respect to w for some $\tilde{N} = \tilde{N}(N, G^2)$. □

Remark

Let us observe that, once coordinates (u, h) chosen, the development as given in (83) is uniquely determined since

$$-k(\lambda) = \lim_{u \to 0} u^2 \ln(h(u)) \tag{85}$$

uniformly in λ.

Proof of proposition 13

In view of lemma 14 and the preceding remark it suffices to prove (81) for (y, A) near some $(y_1, A_1) \in]0, 1/6[\times [-A_0, A_0]$.

The unicity, observed in the remark, will assure that the functions $k_B(y, A)$ and $k_F(y, A)$ locally defined, will give rise to a unique global function.

In the same way, we can find the C^∞-result, by checking the C^ℓ-result (for whatsoever ℓ), in well chosen C^ℓ-coordinates. Let us first treat $k_B(y, A)$.

We take $y_1 \in]0, 1/6[$ and a fixed $\ell \in \mathbb{N}_3$, and we cover the segment in $[n_3, s_2]$, parametrized by $[0, y_1]$, by means of a finite number of open sets $V_1, \ldots, V_{n+1}$ in which X has a C^ℓ-normal form expression as given in proposition 4, i.e.

$$z\frac{\partial}{\partial z} + v^2\frac{\partial}{\partial w} \tag{86}$$

We moreover suppose that in these coordinates $\mathcal{C}_{II} = \cup_A \mathcal{C}_{II}(A)$ is represented by $\{z = 0\}$. For that purpose we could use theorem 3, but we can also observe that for any center manifold $\{z = \varphi(w, v, A)\}$, the change of coordinates

$$(z, w, v, A) \to (z - \varphi(w, v, A), w, v, A)$$

leaves expression (86) invariant.

We want to look at the expression of $\mathcal{C}_I(y, A)$, for (y, A) sufficiently close to (y_1, A_1), within these different neighbourhoods, and especially we want to consider the passage from one neighbourhood to the other. We therefore take sections $T_1, T_2, \ldots, T_n$ transverse to $[n_3, s_2]$ in a way - as represented in figure 23 - that the T_i cut $[n_3, s_2]$ in a monotone decreasing sequence and that each T_i lays in the intersection of the i-th neighbourhood with the $(i + 1)$-th one.

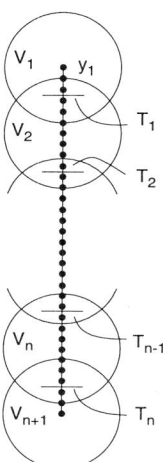

Figure 23
Succession of V_i and T_i along $[0, y_1]$ in $[n_3, s_2]$.

Let in V_1 the section T_1 be represented by $\{w = w_1\}$ with $w_1 < 0$, while the γ_y are given by

$$(z, w, v) = (\alpha(y, v, A), \beta(y, v, A), v) \tag{87}$$

with $\dfrac{\partial \beta}{\partial y}(y, 0, A) > 0$, and α, β both C^ℓ.

As in (61) of lemma 7 we can now calculate the expression $\{z = \Phi_1(y, v, A)\}$ of $\mathcal{C}_I(y, A) \cap T_1$ and get

$$
\begin{aligned}
\Phi_1(y, v, A) &= \alpha(y, v, A) \exp\left(\frac{w_1 - \beta(y, v, A)}{v^2} \right) \\
&= \exp\left(-\frac{\gamma_1(y, A)}{v^2}(1 + \varphi_1(y, v, A)) \right)
\end{aligned} \tag{88}
$$

with $\gamma_1(y, A) = \beta(y, 0, A) - w_1 > 0$, $\dfrac{\partial \gamma_1}{\partial y}(y, A) > 0$ and γ_1 as well as φ_1 are of class C^ℓ; we may also observe that all partial derivatives of φ_1 w.r.t. (y, A) as well as $\dfrac{\partial \varphi_1}{\partial v}$ are $O(v)$ for $v \to 0$, uniformly in (y, A) on any compact subset in the domain of definition.

Before going to V_2 we first look how within V_1 we can pass from the expression $\mathcal{C}_I(y, A) \cap T_1$ to the expression of $\mathcal{C}_I(y, A) \cap \tilde{T}_1$ where $\tilde{T}_1$ represents a section given by $\{w = \delta(z, v, A)\}$ with δ of class C^ℓ and $\delta(0, 0, A) < \overline{w}_1 < 0$ for some $\overline{w}_1$. We parametrize $\tilde{T}_1$ by means of (z, v, A) and represent $\mathcal{C}_I(y, A) \cap \tilde{T}_1$ by $\{z = \tilde{\Phi}_1(y, v, A)\}$:

$$
\begin{aligned}
\tilde{\Phi}_1(y, v, A) &= \Phi_1(y, v, A) \cdot \exp\left(\frac{\delta(\Phi_1(y, v, A), v, A), v, A) - w_1}{v^2}\right) \\
&= \exp\left(-\frac{\tilde{\gamma}_1(y, A)}{v^2}(1 + \tilde{\varphi}_1(y, v, A))\right)
\end{aligned}
\tag{89}
$$

with $\tilde{\gamma}_1(y, A) = \beta(y, 0, A) - \delta(0, 0, A)$.

As such, $\tilde{\gamma}_1$ and $\tilde{\varphi}_1$ still have the same properties as γ_1 and φ_1 in (88).

Within each V_i and for the related normal form coordinates, we will now parametrize T_i by means of (z, v, A) and look at the respective expressions of $\mathcal{C}_I(y, A) \cap T_i = \{z = \Phi_i(y, v, A)\}$

Using lemma 14 (in changing C^ℓ-coordinates on V_i) and performing calculations as in (91) we obtain that

$$
\Phi_i(y, v, A) = \exp\left(-\frac{\gamma_i(y, A)}{v^2}(1 + \varphi_i(y, v, A))\right)
\tag{90}
$$

with $\gamma_i(y, A) > 0$, $\frac{\partial \gamma_i}{\partial y}(y, A) > 0$, γ_i of class C^ℓ and φ_i of class $C^{\ell-2}$ (and of class C^ℓ for $v > 0$); moreover all partial derivatives of φ_i w.r.t. (y, A) as well as $\frac{\partial \varphi_i}{\partial v}$ are $O(v)$ for $v \to 0$, uniformly in (y, A) on any compact subset in the domain of definition.

At the end we arrive in V_{n+1} which is a neighbourhood of s_2, and in which the normal form expression of X (that we take of class $C^{\ell+3}$) is as given in (62), i.e.

$$
z\frac{\partial}{\partial z} + v^2 f(u, v, A)(u\frac{\partial}{\partial u} - v\frac{\partial}{\partial v})
\tag{91}
$$

in some neighbourhood of $s_2 = \{(0, 0, 0, A_0)\}$, f is of class $C^{\ell+3}$. Also here we may suppose that $(\cup_A \mathcal{C}_{II}(A))$ is represented by $\{z = 0\}$.

Inside V_{n+1} we will consider the sections $\tau_1 = \{u = u_1\}$ for some $u_1 > 0$ and $\tau_2 = \{v = v_1\}$ for some $v_1 > 0$.

We know that $\mathcal{C}_I(y, A) \cap \tau_1$ is represented by $\{z = \Phi(y, v, A)\}$ with Φ having properties like in (90) for some γ and φ (similarly defined as the γ_i and φ_i in (90)).

We write $\mathcal{C}_I(y, A) \cap \tau_2 = \{z = \Psi(y, u, A)\}$ and calculate Ψ by means of expression (67) in the proof of lemma 8 :

$$\Psi(y, u, A) = \Phi(y, \frac{uv_1}{u_1}, A) \cdot \exp\left[-\frac{1}{u^2}\left(\frac{u_1}{v_1}\right)^2 \int_{u/u_1}^{1} sF(u_1 s, \frac{uv_1}{su_1}, A)ds\right] \qquad (92)$$

with $F(\alpha, \beta, A) = \dfrac{1}{f(\alpha, \beta, A)} > 0$. The expression $\Phi(u, \dfrac{uv_1}{u_1}, A)$ has similar properties w.r.t. to (y, u, A) as $\Phi(y, v, A)$ has w.r.t. (y, v, A).
We write

$$F(\alpha, \beta, A) = F_0(\alpha, A) + \beta F_1(\alpha, \beta, A) \qquad (93)$$

with F_0 of class $C^{\ell+3}$, F_1 of class $C^{\ell+2}$ and $F_0(\alpha, A) > 0$
Hence

$$sF(u_1 s, \frac{uv_1}{su_1}, A) = sF_0(u_1 s, A) + u\frac{v_1}{u_1}F_1(u_1 s, \frac{uv_1}{su_1}, A) \qquad (94)$$

We see that

$$\int_{u/u_1}^{1} sF_0(u_1 s, A)ds = k_0(A) + \overline{k}_0(u, A) \qquad (95)$$

with $k_0(A) = \int_0^1 sF_0(u_1 s, A)ds > 0$, $k_0(A)$ and $\overline{k}_0(u, A)$ of class $C^{\ell+1}$; $\overline{k}_0(u, A)$, all its derivatives w.r.t. A, as well as $\dfrac{\partial \overline{k}_0}{\partial u}$ are $O(u)$.
The integral of the second term in (94) and all its derivatives w.r.t. A are clearly $O(u)$; let us now show that its first derivative w.r.t. u stays bounded; (in fact the integral is C^1).
For $u > 0$ we get :

$$\frac{\partial}{\partial u}\left(v \int_{u/u_1}^{1} F_1(u_1 s, \frac{uv_1}{su_1}, A)ds\right) =$$
$$\int_{u/u_1}^{1} F_1(u_1 s, \frac{uv_1}{su_1}, A)ds - \frac{u}{u_1}F_1(u, v_1, A) + \frac{uv_1}{u_1}\int_{u/u_1}^{1}\frac{1}{s}\frac{\partial F_1}{\partial \beta}(u_1 s, \frac{uv_1}{su_1}, A)ds \qquad (96)$$

For the first integral in (96) we write

$$F_1(\alpha, \beta, A) = F_1^0(\alpha, A) + \beta F_1^1(a, A) + \beta^2 F_1^2(\alpha, \beta, A) \qquad (97)$$

The F_i^j are of class $C^{\ell+2-j}$ and hence

$$\int_{u/u_1}^{1} F_1^0(u_1 s, A) ds$$

is of class $C^{\ell+2}$.

If we write $F_1^1(\alpha, A) = F_2^1(0, A) + \alpha F_3^1(\alpha, A)$, we see that

$$u \frac{v_1}{u_1} \int_{u/u_1}^{1} \frac{1}{s} F_1^1(\alpha, A) ds = k_1(A) u \ln u + u \overline{k}_1(u, A) \tag{98}$$

with k_1 and $\overline{k}_1$ both at least $C^{\ell+1}$.

At last

$$u^2 \left(\frac{v_1}{u_1}\right)^2 \left| \int_{u/u_1}^{1} \frac{1}{s^2} F_1^2\left(u_1 s, \frac{u v_1}{s u_1}, A\right) ds \right| \le Cu \tag{99}$$

for some constant C, uniformly in A. As the expression is surely C^0 for $u > 0$, it has to be C^0 everywhere.

The third integral in (96) can be treated in the same way as the first, and the second is clearly $C^{\ell+2}$.

We can conclude from (99) and the calculations for the integral in it , that Ψ satisfies an expression as in (96) with respect to u instead of v.

Because of lemma 14, and using the flow box theorem; we hence obtain that a similar result is valid for $\mathcal{C}_{II}(A) \cap T$ and $\mathcal{C}_I(y, A) \cap T$ as required.

This finishes the proof for $k_B(y, A)$.

The proof for $k_F(y, A)$ is completely analogous. $\square$

With methods similar to the ones used in the proof we can also prove a few results on $k_B(y, A)$ and $k_F(y, A)$ for $y = 0, 1/6$. However a much better result can be obtained with completely different techniques.

Theorem 15 : *With the notations of proposition 13 and for $A_0 > 0$ sufficiently small we have :*

$$k_B(y, A) = \int_0^{x_0(y)} x(1+x)^2 dx$$

(100)

$$k_F(y, A) = \int_0^{x_1(y)} x(1+x)^2 dx$$

where $x_0(y)$ and $x_1(y)$ respectively are the largest negative and the unique positive solution of

$$\frac{x^2}{2} + \frac{x^3}{3} = y,$$

for $y \in]0, 1/6[$.

Remark

1. We see that $k_B(y, A)$ and $k_F(y, A)$ are independent of A and they both have a limit for y tending to 0 or $1/6$. Let us write :

$$k_B(y) = k_B(y, A)$$
$$k_F(y) = k_F(y, A)$$

(101)

for $y \in [0, 1/6]$, by making the continuous extension :

$$k_B(0) = k_F(0) = 0$$
$$k_B(1/6) = 1/12, \quad k_F(1/6) = \frac{43}{192}.$$

(102)

We easily see that

$$\frac{\partial k_B}{\partial y}(y) = \left[x(1+x)^2 x' \right]_0^{x_0(y)} = 1 + x_0(y)$$

(103)

since $(x + x^2)x' = 1$ for $x = x_0(y)$.
Similarly

$$\frac{\partial k_F}{\partial y}(y) = 1 + x_1(y)$$

(104)

As such we obtain that k_B and k_F extend in a C^1-way on $[0, 1/6]$ and

$$k_F(y) > k_B(y)$$

(105)

for $y \in]0, 1/6]$.

2. The function $k(y)$ as used in theorem 1 can be defined as :

$$k(y) = k_B(y) \tag{106}$$

Proof of theorem 15 :

We give the proof for k_B, the one for k_F being similar. Let us consider the section $\mathcal{G}_I$, from 3.3.1, near $(0, y)$ and consider the backward "transition map" from $\mathcal{G}_I$ to T, that we denote by B, i.e.

$$(u, h, A) = (B_1(v, y, A), \ B_2(v, y, A), A) \tag{107}$$

To each point (v, y, A) in $\mathcal{G}_I$, with $0 < y < 1/6$ and $A \in [-A_0, A_0]$ for $A_0 > 0$ and sufficiently small corresponds an orbit $O_{(v,y,A)}$ that stays within the leaf parametrized by (v, A). We can also parametrize this leaf by (u, A). As such, from proposition 13, we obtain that

$$B_2(u, y, A) = B_{II}(u, A) - exp\left(-\frac{k_B(y, A)}{u^2}(1 + \varphi_B(u, y, A))\right)$$

and because of the properties proved in that proposition we get

$$\frac{\partial B_2}{\partial y}(u, y, A) = \frac{1}{u^2} \cdot \frac{\partial k_B}{\partial y}(y, A)(1 + O(u))exp\left(-\frac{k_B(y, A)}{u^2}(1 + O(u))\right)$$

and even

$$\frac{\partial B_2}{\partial y}(u, y, A) = \frac{1}{u^2} \cdot exp\left(-\frac{k_B(y, A)}{u^2}(1 + O(u))\right) \tag{108}$$

uniformly in (y, A) on compacta in $]0, 1/6[\times[-A_0, A_0]$ for $A_0 > 0$ sufficiently small. Within the leaf parametrized by (u, A) and for the orbit $O_{(u,y,A)}$ there is another expression for $\partial B_2/\partial y$ as given in [ALGM], and that we obtain by working with $X = X_{\varepsilon,a}$ for $\varepsilon = u^2$ and $a = uA$, nl :

$$\frac{\partial B_2}{\partial y}(u, y, A) = \frac{\alpha(u, y, A)}{u^2} \cdot exp\left(\int_0^{T(u,y)} div \ X(O_{(u,y,A)}(t))dt\right) \tag{109}$$

where $O_{(u,y,A)}$ denotes the orbit, $T(u, y, A)$ the time it needs go from $\mathcal{G}_I$ to T, and $\alpha(u, y, A)$ is some strictly positive C^∞ function.

Now $div\ X_{\varepsilon,a} = -(x + x^2)$ and if we write $dt = \dfrac{dy'}{\varepsilon(a-x)}$ (using y' instead of y in order not to confuse with the parameter y) we get

$$\int_0^{T(u,y,A)} div\ X(O_{(u,y,A)}(t))dt = -\frac{1}{u^2} \int_{O_{(u,y,A)}} \frac{(x+x^2)}{uA-x}dy' \tag{110}$$

From (108), (109) and (110) follows that

$$k_B(y,A) = \lim_{u \to 0} \int_{O_{(u,y,A)}} \frac{(x+x^2)}{uA-x}dy' \tag{111}$$

uniformly in (y,A) on compacta in $]0,1/6[\times[-A_0,A_0]$ for A_0 sufficiently small. Let us now show that the limit in (111) is equal to

$$-\int_{{}^L\Gamma_y^I}(1+x)dy' = \int_0^{x_0(y)} x(1+x)^2 dx$$

where ${}^L\Gamma_y^I$ denotes the part of (the l.p.s.) Γ_y^I in $\{x \le 0\}$ and followed in negative time.

It clearly suffices to prove this result for each $(y,A) \in]0,1/6[\times [-A_0,A_0]$ separately. We recall that

$$\int_{O_{(u,y,A)}} \frac{x+x^2}{uA-x}dy' = -u^2 \int_0^{T(u,y,A)} div\ X(O_{(u,y,A)}(t))dt \tag{112}$$

The integrandum of the first integral is regular near the horizontal part $\{y' = y\}$ of ${}^L\Gamma_y^I$ for $x < -x_0$ as well as near the curve $\{y' = x^2/2 + x^3/3\}$ for $x < -x_0$ with $x_0 > 0$ some small value.

The same piece of ${}^L\Gamma_y^I$ that we will denote by $E_y^{x_o}$ is approached by $O_{(u,y,A)}$ in a C^1 way, except at the point $\{(x^-,y)\}$, solution of $\{x^2/2 + x^3/3 = y\}$ with $x^- < 0$.

Near that point we can use the normal forms as given in proposition 4 and the calculations as made in 3.2.1. Let us write $E_{(u,y,A)}^{x_o}$ for the piece of $O_{(u,y,A)}$ inside $\{x \le -x_0\}$.

It is now an easy exercise to show that

$$\int_{E_{(u,y,A)}^{x_0}} \frac{x+x^2}{uA-x}dy' \to -\int_{E_y^{x_0}}(1+x)dy'$$

It is also clear that the part of the integral along $O_{(u,y,A)}$ for $-x_0 \leq x \leq 0$ and near the "horizontal" piece of Γ_y^I tends to zero for $u \to 0$, by considering the second integral of (112), observing that the divergence and the time are bounded when $u \to 0$.

For the rest of the integral (in (112)) we make a subdivision in two pieces, by taking $x \in [-x_0, -ru]$ or $x \in [-ru, 0]$, with $r > 0$ some fixed number.

We can show that the part of the integral related to $x \in [-ru, 0]$ tends to zero for $u \to 0$.

To that end we use the second expression in (112).

The divergence is a bounded function and we can estimate the time by using the family rescaling as given in 2.2.1. After the rescaling, the new time - on the chosen piece of $O_{(u,y,A)}$ is uniformly bounded for $u \to 0$, implying that the original time is of order $O(1/u)$ for $u \to 0$, uniformly on $\{-ru \leq x \leq 0\}$.

The u^2-factor in front of the (second) integral implies the result.

We are left with the integral $\int_{-x_0}^{-ru} \dfrac{x + x^2}{uA - x} dy'$, that we write as

$$\int_{ru}^{x_0} \frac{x f(u,x)}{uA - x} dx \tag{113}$$

with $f(u,x)$ some function, of class C^0 for each u separately, and uniformly bounded on $\{u \geq 0, u \leq x \leq x_0\}$ - for a proof see further -, and we take $A \in [-A_0, A_0]$ with $A_0 > 0$ sufficiently small.

If we introduce the change of variables $x = ru + t(x_0 - ru)$ then (113) changes into

$$(ru - x_0) \int_0^1 \frac{(ru + t(x_0 - ru)) f(u, ru + t(x_0 - ru))}{ru - uA + t(x_0 - ru)} dt \tag{114}$$

If we use

$$(ru + t(x_0 - ru)) \mid f(u, ru + t(x_0 - ru) \mid \leq C(ru + tx_0)$$

for some constant $C > 0$ and

$$ru - uA + t(x_0 - ru) \geq \frac{ru}{2} + t\frac{x_0}{2}$$

for $A_0 \leq r/2$ and $ru \leq x_0/2$, we see that the integrandum of (114) is bounded by a constant. The theorem of Lebesgue on the dominated convergence now implies that

(114) tends to $-\int\limits_0^{x_0} f(0,x)dx$, what we wanted to prove.

There only remains to check that the function $f(u,x)$ in (113) has the required properties. This clearly follows if the slope of the tangent field of $O_{(u,y,A)}$ is uniformly bounded on $\{u > 0, -x_0 \geq x \geq -u\}$. The expression of the slope is

$$\frac{\varepsilon(a-x)}{y - \dfrac{x^2}{2} - \dfrac{x^3}{3}} \tag{115}$$

for $\varepsilon > 0$ (while it is $-(x+x^2)$ for $\varepsilon = 0$).

To get the result we write expression (115) in chart FR1, see (28), using $(x,y,\varepsilon,a) = (u\overline{x}, u^2, u^2v^2, uvA)$; this gives

$$\frac{uv^2(vA - \overline{x})}{1 - \dfrac{\overline{x}^2}{2} - \dfrac{u}{3}\overline{x}^3}$$

The pieces of $O_{(u,y,A)}$ under consideration lay in the center manifolds $C_I(y,A)$ of $(\overline{x}, u, v) = (-\sqrt{2}, 0, 0)$, going from $\{u\overline{x} = -x_0\}$ to $\{\overline{x} = rv\}$, (corresponding respectively to $\{x = -x_0\}$ and $\{x = -ru = -r\sqrt{\varepsilon}\}$).

The result will now follow by showing that for v and u sufficiently small and for orbits staying in $C_I(y,A)$ we have :

$$-g(u,\overline{x}) = -1 + \frac{\overline{x}^2}{2} + \frac{u}{3}\overline{x}^3 \geq Dv^2 \tag{116}$$

for some $D > 0$.

From expression (28) it is clear that the surfaces $\{g = 0\}$ and $C_I(y,A)$ are tangent along the line $\{v = 0, \ g = 0\} = C_I(y,A) \cap \{v = 0\}$.

To have an idea on their contact, we consider their intersection with $\{u = 0\}$:
$\{g = u = 0\} = \{\overline{x} = -\sqrt{2}\}$ while a standard formal calculation shows that $C_I(y,A) \cap \{u = 0\}$ is given by

$$\overline{x} = -\sqrt{2} - \frac{1}{\sqrt{2}}v^2 + \dots$$

This clearly induces (116) for $D = \dfrac{1}{2\sqrt{2}}$, at least if we keep u and v sufficiently small. This last restriction is a consequence of a good choice of x_0 and of r, e.g. we have to take x_0 small enough and r large enough.   □

3.3.2   Foliation of type II

We proceed like in the previous section and as initial condition we choose a C^∞ family of segments γ, foliating a rectangle $\mathcal{G}_{II} \subset P$ defined in the following way (see figure 24).

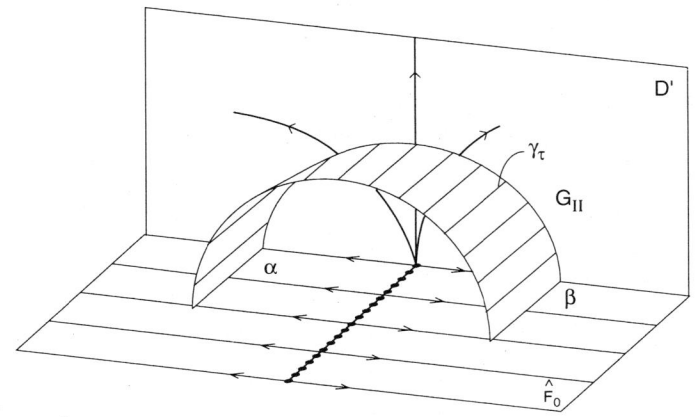

Figure 24
$\mathcal{G}_{II} \subset P$, *foliated by segments* γ_θ.

One side of $\mathcal{G}_{II}$ is a circle-arc $[\alpha, \beta]$ in D' near n_3 transverse to the orbits in D', with $\alpha, \beta \in \sigma_2$ on different sides of n_3. The segments γ are transversally cutting D' in points of $[\alpha, \beta]$ and we can parametrize them by taking an angular variable $\theta \in [0, \pi]$, representing $[\alpha, \beta]$ in a sense that $\alpha = \{\theta = 0\}$ and $\beta = \{\theta = \pi\}$. γ_θ stands for the segment ending at the point on $[\alpha, \beta]$ with parameter θ.

Like in 3.3.1 we define the foliation $\mathcal{C}_{II}(\theta, A)$ by saturating the γ_θ (see also 3.2.2). It will be defined for $A \in [-A_0, A_0]$ with $A_0 > 0$ sufficiently small and $\theta \in]0, \pi[$.

Each leaf $\mathcal{C}_{II}(\theta, A)$ cuts T in negative time along a C^∞ curve $\{h = B_{II}(u, \theta, A)\}$ and in positive time along a C^∞ curve $\{h = F_{II}(u, \theta, A)\}$. The domain of definition is

$$\mathcal{D}_{II} = \{(u, \theta, A) | 0 \le u \le u_{II}, 0 < \theta < \pi, -A_0 \le A \le A_0\} \qquad (117)$$

where u_{II} is some strictly positive constant.

The reference manifold $\mathcal{C}_{II}(A)$, used in 3.3.1, can e.g. be chosen to be

$$\mathcal{C}_{II}(A) = \mathcal{C}_{II}(\pi/2, A) \qquad (118)$$

and hence

$$F_{II}(u, A) = F_{II}(u, \pi/2, A)$$
$$B_{II}(u, A) = B_{II}(u, \pi/2, A) \tag{119}$$

Proposition 16

In terms of the notations introduced above one obtains that B_{II} and F_{II} are C^∞ on $\mathcal{D}_{II} \cap \{u > 0\}$ while for all $0 < K < k_B(1/6)$ there is a sufficiently small $u_{II} > 0$ such that on $\mathcal{D}_{II} \cap \{0 < u \le u_{II}\}$:

$$\left| \frac{\partial^{i+j+k}}{\partial u^i \partial \theta^j \partial A^k} \left(B_{II}(u, \theta, A) - B_{II}(u, \pi/2, A) \right) \right| \le exp\left(-\frac{K}{u^2} \right) \tag{120}$$

for $0 \le i + j + k \le 1$.

Similarly for all $0 < K' < k_F(1/6)$ there is a sufficiently small $u_{II} > 0$ such that on $\mathcal{D}_{II} \cap \{0 < u \le u_{II}\}$:

$$\left| \frac{\partial^{i+j+k}}{\partial u^i \partial \theta^j \partial A^k} \left(F_{II}(u, \theta, A) - F_{II}(u, \pi/2, A) \right) \right| \le exp\left(-\frac{K'}{u^2} \right) \tag{121}$$

for $0 \le i + j + k \le 1$.

Proof

Let us start with B_{II}. We work similarly as in the proof of proposition 13, except for the fact that we now also need to pass near the point n_3, in the neighbourhood of which we can start the construction.

Let us take a neighbourhood U of n_3 in which $-X$ has a C^ℓ-normal form (for equivalence) as in (71) of lemma 10, nl.

$$u\frac{\partial}{\partial u} - 3v\frac{\partial}{\partial v} - \frac{F(u, v, A)}{v^2} z\frac{\partial}{\partial z}$$

with $F(u, v, A) > 0$.

Let us fix $z = z_1 > 0$ (the case $z = z_1 < 0$ being similar) and let us choose segments $\gamma_{\tilde{v}} = \{(u, v, z) = (\tilde{u}, V(\tilde{u}, \tilde{v}, A), z_1)\}$ with $v(0, \tilde{v}, A) = \tilde{v}$, $v(\tilde{u}, 0, A) = 0$, $0 \le \tilde{u} \le u_0$, $u_0 > 0$, and the $\gamma_{\tilde{v}}$ being a C^∞-foliation of

$$\{(u, v, z_1) \mid 0 \le u \le u_0, \ 0 \le v \le V(u, \tilde{v}_0, A)\} = S_A$$

with $\tilde{v}_0 > 0$ and all S_A lying inside U.

Let us also consider the segment $C_{u_1} = \{(u, v, z) = (u_1, \tilde{v}, z_1)\}$ with $u_0 < u_1$. We now

look how the saturations of C_{u_1} and of $\gamma_{\tilde{v}}$ cut the plane $\{u = u_2\}$ for $u_1 < u_2$.
Let us denote the first intersection by $\{z = \Phi(v, A)\}$ and the others by $\{z = \Psi(v, \tilde{v}, A)\}$.

We have

$$\Phi(v, A) = z_1 exp\left[-\left(\frac{u_1}{u_2}\right)^6 \frac{1}{v^2} \int\limits_0^{\ln u_2/u_1} e^{6\tau} F(u_1 e^\tau, \left(\frac{u_2}{u_1}\right)^3 v e^{-3\tau}, A) d\tau\right]$$

and taking $s = e^\tau$ we get

$$\Phi(v, A) = z_1 exp\left[-\left(\frac{u_1}{u_2}\right)^6 \frac{1}{v^2} \int\limits_1^{u_2/u_1} s^5 F(u_1 s, \left(\frac{u_2}{u_1}\right)^3 \frac{v}{s^3}, A) ds\right] \qquad (122)$$

We will now develop

$$F(\alpha, \beta, A) = F_0(A) + \alpha F_1(\alpha, A) + \beta F_2(\alpha, \beta, A) \qquad (123)$$

inducing a decomposition

$$\left(\frac{u_1}{u_2}\right)^6 \frac{1}{v^2} \int\limits_1^{u_2/u_1} s^5 F(u_1 s, \left(\frac{u_2}{u_1}\right)^3 \frac{v}{s^3}, A) ds = I_1 + I_2 + I_3 \qquad (124)$$

with

$$I_1 = \left(\frac{u_1}{u_2}\right)^6 \frac{1}{v^2} F_0(A) \int\limits_1^{u_2/u_1} s^5 ds = \frac{F_0(A)}{6}\left(1 - \frac{u_1^6}{u_2^6}\right)\frac{1}{v^2} \qquad (125)$$

$$| I_2 | \leq \frac{u_1^7}{u_2^6} \cdot \max_U | F_1 | \cdot \frac{1}{v^2} \int\limits_1^{u_2/u_1} s^6 ds = \frac{1}{7} \cdot \max_U | F_1 | \cdot \frac{1}{v^2}\left(u_2 - \frac{u_1^7}{u_2^6}\right)$$

$$\leq \frac{1}{7} \cdot \max_U | F_1 | \cdot u_2 \frac{1}{v^2} \qquad (126)$$

$$| I_3 | \leq \left(\frac{u_1}{u_2}\right)^3 \cdot \max_U | F_2 | \cdot \frac{1}{v} \int\limits_1^{u_2/u_1} s^2 ds = \frac{1}{3} \cdot \max_U | F_2 | \cdot \frac{1}{v}\left(1 - \left(\frac{u_1}{u_2}\right)^3\right)$$

$$\leq \frac{1}{3} \cdot \max_U | F_2 | \cdot \frac{1}{v} \qquad (127)$$

Choosing u_2 as well as u_1/u_2 sufficiently small we see from (124) - (127) that there exists some $\alpha(A) > 0$ close to $\dfrac{F_0(A)}{6}$, such that

$$\Phi(v, A) \geq e^{-\frac{\alpha(A)}{v^2}} \tag{128}$$

Let us now look at $\Psi(v, \tilde{v}, A)$ but restrict to the special case that $V(\tilde{u}, \tilde{v}, A) = \tilde{v}$:

$$\Psi^S(v, \tilde{v}, A) = z_1 exp\left[-\frac{1}{\tilde{v}^2} \int\limits_0^{\frac{1}{3}\ln(\tilde{v}/v)} e^{6\tau} F(u_2 \left(\frac{v}{\tilde{v}}\right)^{1/3} e^\tau, \tilde{v}e^{-3\tau}, A)d\tau \right]$$

and taking $s = e^{3\tau}$ we get

$$\Psi^S(v, \tilde{v}, A) = z_1 exp\left[-\frac{1}{3}\frac{1}{\tilde{v}^2} \int\limits_1^{\tilde{v}/v} sF(u_2 \left(\frac{v}{\tilde{v}}s\right)^{1/3}, \tilde{v}s^{-3}, A)ds \right] \tag{129}$$

The development (123) for F induces that

$$\frac{1}{3}\frac{1}{\tilde{v}^2} \int\limits_1^{\tilde{v}/v} sF\left(u_2 \left(\frac{v}{\tilde{v}}s\right)^{1/3}, \tilde{v}s^{-3}, A\right) ds = J_1 + J_2 + J_3 \tag{130}$$

with

$$J_1 = \frac{1}{3}\frac{1}{\tilde{v}^2}F_0(A) \int\limits_1^{\tilde{v}/v} s \, ds = \frac{F_0(A)}{6}\left(1 - \frac{u_0^6}{u_2^6}\right) \cdot \frac{1}{v^2} \tag{131}$$

$$| J_2 | \leq \frac{u_2}{3} \cdot \max_U | F_1 | \cdot \left(\frac{v}{\tilde{v}}\right)^{1/3} \int\limits_1^{\tilde{v}/v} s^{4/3}ds = \frac{u_2}{7} \cdot \max_U | F_1 | \cdot \left(\frac{\tilde{v}}{v^2} - \left(\frac{v}{\tilde{v}}\right)^{1/3}\right)$$

$$\leq \frac{u_2}{3} \cdot \max_U | F_1 | \cdot \tilde{v} \cdot \frac{1}{v^2} \tag{132}$$

$$| J_3 | \leq \frac{1}{3}\frac{1}{\tilde{v}} \cdot \max_U | F_2 | \cdot \int\limits_1^{\tilde{v}/v} s^{-2}ds = \frac{1}{3} \cdot \max_U | F_2 | \cdot \frac{1}{\tilde{v}}\left(1 - \frac{v}{\tilde{v}}\right)$$

$$\leq \frac{1}{3} \cdot \max_U | F_1 | \cdot \left(\frac{u_0}{u_2}\right)^3 \frac{1}{v} \tag{133}$$

At certain steps in the reasoning we use the fact that $u_2^3 v = \tilde{u}^3 \tilde{v}$ with $| \tilde{u} | \leq u_0$. Choosing again u_2 small and taking $u_0 < u_1$ we find some $\beta(A) > 0$, such that

$$0 \leq \Psi^S(v, \tilde{v}, A) \leq e^{-\frac{\beta(A)}{v^2}} \tag{134}$$

Comparing (125) and (131) it is clear that we can choose $\alpha(A) < \beta(A)$.

The elaboration for the general $V(\tilde{u}, \tilde{v}, A)$ leads to the same inequality (134).

The inequality is also true for $z_1 < 0$ and remains correct if we complete the foliation of segments $\gamma_{\tilde{v}} \subset \{z_1 = 0\}$ by segments $\gamma_{\tilde{z}} \subset \{v_1 = 0\}$ for some $v_1 > 0$ sufficiently small; for this last statement we refer to expression (74) in lemma 10.

Now (128) and (134) - using $\alpha(A) < \beta(A)$ - will induce result (120), for $i + j + k = 0$, once the center manifolds determined by $\gamma_{\tilde{v}}$ and C_{u_1} arrive at T.

In order to obtain expression (120) for $i + j + k = 1$, there is - within U - some technical work to do of the kind as we did in the proof of proposition 13.

These calculation lead to similar inequalities as (128) and (134) inducing the required result.

The statements in (121) are easier to prove and follow from the proof of proposition 13, combined with lemma 11. $\square$

3.3.3   Foliations of type III

We proceed like in 3.3.1 and as initial conditions we choose a C^∞ family of segments γ, foliating a rectangle $\mathcal{G}_{III} \subset P$ defined in the following way (see figure 25) :

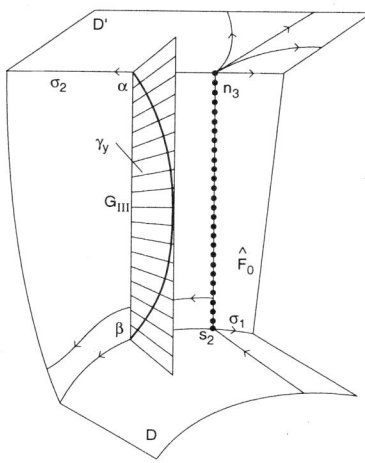

Figure 25

$\mathcal{G}_{III} \subset P$, foliated by segments γ_y.

We again call $\mathcal{G}_{III} \cap \hat{F}_0 = [\alpha, \beta]$ with $\mathcal{G}_{III} \cap \sigma_1 = \{\beta\}$ and $\mathcal{G}_{III} \cap \sigma_2 = \{\alpha\}$, and parametrize $[\alpha, \beta]$ by $y \in [0, 1/6]$. But this time $[\alpha, \beta]$ is to the "left" of $[n_3, s_2]$ (see figure 25).

$\mathcal{G}_{III}$ again has sides on D and D', and $\tilde{\gamma}_y$ denotes the segment cutting $\hat{F}_0$ at a point with parameter y.

Similarly as in 3.3.1 we can define $\mathcal{C}_{III}(y, A)$, as well as $F_{III}(u, y, A)$ and $B_{III}(u, y, A)$. Let us specify that $\{h = F_{III}(u, y, A)\}$ and $\{h = B_{III}(u, y, A)\}$ denote the intersections of $\mathcal{C}_{III}(y, A)$ with T in respectively positive and negative time. F_{III} and B_{III} are defined on

$$\mathcal{D}_{III} = \{(u, y, A) \mid -A_0 \leq A \leq A_0, 0 \leq y \leq 1/6, 0 \leq u \leq u_{III}(y)\} \qquad (135)$$

for some continuous function $u_{III}(y)$ with $u_{III}(y) > 0$ for $0 < y < 1/6$ and $u_{III}(0) = u_{III}(1/6) = 0$. Similar techniques as the ones used in the proof of the propositions 13 and 16 - taking into account theorem 15 - permit to show the following result.

Proposition 17

In terms of the notations introduced above one obtains that B_{III} and F_{III} are C^∞ on $\mathcal{D}_{III} \cap \{u > 0\}$. Moreover :

i)

$$B_{III}(u, y, A) - B_{II}(u, A) = exp\left(-\frac{k_B(y)}{u^2}(1 + \tilde{\varphi}_B(u, y, A))\right) \qquad (136)$$

where $\tilde{\varphi}_B$ is C^∞ on $\mathcal{D}_{III} \cap \{u > 0\}$ for u_{III} sufficiently small, and all partial derivatives of $\tilde{\varphi}_B$ with respect to (y, A) as well as $u\dfrac{\partial \tilde{\varphi}_B}{\partial u}(u, y, A)$ are $O(u)$ for $u \to 0$, uniformly in (y, A) on any compact subset of $]0, 1/6[\times [-A_0, A_0]$.

ii) for all $0 < K' < k_F(1/6)$ there is a sufficiently small u_{III} such that on $\mathcal{D}_{III} \cap \{0 < u \le u_{III}\}$ one has

$$\left|\frac{\partial^{i+j+k}}{\partial u^i \partial \theta^j \partial A^k}(F_{III}(u, y, A) - F_{II}(u, A))\right| \le exp\left(-\frac{K'}{u^2}\right) \qquad (137)$$

for $0 \le i + j + k \le 1$.

4 The canard phenomenon

We are now finally going to study how the different limit periodic sets are approached by limit cycles. We first start with the "small" l.p.s., namely the point s itself. Next we will tackle the l.p.s. of respectively type I, II and III.

4.1 The small limit periodic set

If a sequence of limit cycles $\{\gamma_i\}_i$ tends to $\{s\}$, we see it in the blown-up space M as a sequence $\{\hat{\gamma}_i\}$. Taking a subsequence - if necessary - we may suppose that $\{\hat{\gamma}_i\}_i$ is convergent. It tends to a l.p.s. in Σ_1 which has necessarily to be a subset of $P_{(1,0)}$. It hence is a periodic orbit around the center, or the center itself, or the connection Γ completed at infinity (see figure 7) for $\overline{a} = 0$, i.e. $A = 0$.

In the neighbourhood of the center and near each periodic orbit the local field X is represented by the family $\overline{X}_1$ of (9) written in chart FR1 ($\overline{\varepsilon} = 1$), and depending on the parameters $(u, \overline{a})$. The vector field of $\overline{X}_1$ for $(u, \overline{a}) = (0, 0)$ is a symmetric center and the limit cycles of the systems for $(u, \overline{a})$ near $(0, 0)$ can be studied by the usual perturbation techniques (e.g. see [ALGM]).

To that end, we consider

$$X_S = \overline{X}_{1,(1,0)} = \overline{X}_1 \mid \overline{D}_{(1,0)} = (\overline{y} - \frac{\overline{x}^2}{2})\frac{\partial}{\partial \overline{x}} - \overline{x}\frac{\partial}{\partial \overline{y}} \tag{138}$$

It is a time-reversible vector field and its dual 1-form is

$$\omega_S = (\overline{y} - \frac{\overline{x}^2}{2})d\overline{y} + \overline{x}d\overline{x} \tag{139}$$

If we consider

$$H(\overline{x}, \overline{y}) = -e^{-\overline{y}}(\overline{y} - \frac{\overline{x}^2}{2} + 1) \tag{140}$$

then we get

$$\omega_S = e^{\overline{y}}dH \tag{141}$$

System (138) is hence integrable with integrating factor $e^{-\overline{y}}$ and H from (140) as first integral.

Of course the integrating factor is not unique, but $e^{-\overline{y}}$ has the advantage that the related Hamiltonian H is zero on the connection $\{\overline{y} = \dfrac{\overline{x}^2}{2} - 1\}$, and hence also at infinity, while $e^{-\overline{y}}$ is flat for $\overline{y} = \infty$.

Let us now take the dual 1-form of the family $\overline{X}_1$:

$$\omega_{u,\overline{a}}(\overline{x},\overline{y}) = \omega_S(\overline{x},\overline{y}) - \overline{a}d\overline{x} - \frac{u}{3}\overline{x}^3 d\overline{y} \tag{142}$$

It is equivalent to

$$e^{-\overline{y}}\omega_{u,\overline{a}}(\overline{x},\overline{y}) = dH(\overline{x},\overline{y}) - e^{-\overline{y}}(\overline{a}d\overline{x} + \frac{u}{3}\overline{x}^3 d\overline{y}) \tag{143}$$

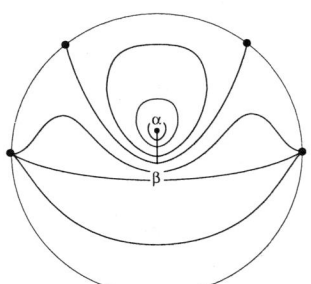

Figure 26
Compactification of X_S.

We consider the Poincaré-mapping $P_{u,\overline{a}}$ of $\overline{X}_1$ with respect to a transverse section $[\alpha,\beta]$ joining the center $\alpha = (0,0)$ to the point $\beta = (0,-1)$ on Γ, along the $\overline{y}$-axis (see figure 26). We parametrize $[\alpha,\beta]$ by the value h of the Hamiltonian. The asymptotic development of $P_{u,\overline{a}}$ is given by

$$P_{u,\overline{a}}(h) = h + \overline{a}I_1(h) + uI_2(h) + o(|\,\overline{a},u\,|) \tag{144}$$

where

$$I_1(h) = \int\limits_{\Gamma_h} e^{-\overline{y}}d\overline{x}, \qquad I_2(h) = \frac{1}{3}\int\limits_{\Gamma_h} e^{-\overline{y}}\overline{x}^3 d\overline{y} \tag{145}$$

with $\Gamma_h = H^{-1}(h)$ and $h \in\,]0,-1[$, Γ_{-1} represents the center while Γ_0 represents the connection.

As in the theory of the Bogdanov-integral ([B]) the bifurcations for small $(u,\overline{a})$ are completely determined by the integrals I_1 and I_2 and rely on the following result.

Theorem 18

The quotient $I_2(h)/I_1(h)$ has a strictly positive derivative with respect to h on $]-1,0[$; $I_2(0)/I_1(0) = 1$ and $I_2(h)/I_1(h) \cong \dfrac{1}{2}(h + 1)$ for $h \to -1$.

Proof : The statement on the derivative can be found in [vG]. The results on the asymptotics for $h \to -1$, easily follow from an asymptotic calculation on the integrals for $(\overline{x}, \overline{y})$ near $(0,0)$.

Let us now calculate $I_2(0)/I_1(0)$. The closed orbits Γ_h tend to the polycycle $\Gamma_0 = [s_2, s_3] \cup \Gamma$, where $[s_2, s_3]$ is defined in 2.2.2. As the integrands of I_1 and I_2 are zero at infinity on the segment $[s_2, s_3]$ in σ_1 we get

$$I_1(0) = \int_\Gamma e^{-\overline{y}} d\overline{x}, \qquad I_2(0) = \frac{1}{3} \int_\Gamma e^{-\overline{y}} \overline{x}^3 d\overline{y} \tag{146}$$

with $\Gamma = \{\overline{y} = \dfrac{1}{2}\overline{x}^2 - 1\}$.
Knowing that

$$I_1(0) = -e \cdot \int_{-\infty}^{\infty} e^{-(1/2)\overline{x}^2} d\overline{x}, \qquad I_2(0) = -\frac{1}{3} e \cdot \int_{-\infty}^{\infty} e^{-(1/2)\overline{x}^2} \overline{x}^4 d\overline{x} \tag{147}$$

we get $I_2(0) = I_1(0)$ by integrating by parts. $\qquad \square$

As a consequence of this theorem and of the general theory on perturbations from a Hamiltonian vector field we get that on $\overline{D} = \{h \leq h_{\max}\}$, for some $-1 < h_{\max} < 0$, the equation of fixed points

$$P_{u,\overline{a}}(h) - h = \overline{a} I_1(h) + u I_2(h) + o(|\overline{a}, u|) = 0 \tag{148}$$

is a C^∞ perturbation of the equation :

$$\overline{a} I_1(h) + u I_2(h) = 0 \tag{149}$$

This equation is regular in the sense that it has a unique and non-degenerate solution $h(\overline{a}/u)$ for each value $\overline{a}/u$ between 0 and $-I_2(h_{\max})/I_1(h_{\max})$.

As a result, in the parameter space $(u, \overline{a})$ we have a curve of generic Hopf-bifurcation, tangent at $(0,0)$ to the line $\{\overline{a} = 0\}$ (In fact it is $\{\overline{a} = 0\}$ itself). Moreover the cycle born at the Hopf-bifurcation stays hyperbolic and grows monotonically until getting, for $h = h_{\max}$, in a position close to $\Gamma_{h_{\max}}$, at least along a curve, tangent at $(0,0)$ to the line $\{\overline{a} I_1(h_{\max}) + u I_2(h_{\max}) = 0\}$.

We can of course take $h_{\max}$ as close to 0 as we want but not $h_{\max} = 0$ as formula (148) does not permit to conclude when $h \to 0$. The reason is twofold : Γ_0 has singular points, and the local field X cannot be represented by $\overline{X}_1$ near Γ_0. We can use $\overline{X}_1$ near any compact segment of $\Gamma \subset \Gamma_0$, but near $[s_2, s_3] \subset \Gamma_0$ we have to use phase-directional rescaling, and hence e.g. chart PR1 (see (24)).

If we compactify in this way $\overline{X}_{1,(1,0)}$, then formula (144) for $u = h = 0$ gives a development of the separation $\Delta(\overline{a})$ of the center manifolds $\Gamma_{\overline{a}}^+$ and $\Gamma_{\overline{a}}^-$ at respectively s_3 and s_2, measured in parameter h, along the $\overline{y}$-axis

$$\Delta(\overline{a}) = \overline{a} I_1(0) + o(\overline{a}) \qquad (150)$$

where $\Delta(\overline{a}) = h(\Gamma_{\overline{a}}^+ \cap [\alpha, \beta]) - h(\Gamma_{\overline{a}}^- \cap [\alpha, \beta])$ (see figure 27).

The way to prove this is to take the general well known formula for regular points x (see figure 27) and to look at the limit for x tending to $x_0 \in]s_2, s_3[$ on σ_1.

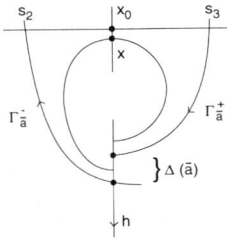

Figure 27

Separation of the center manifold in Σ_1.

Since $I_1(0) < 0$, we see that $\Delta(\overline{a}) < 0$ for $\overline{a} > 0$, corresponding to the pictures in figure 7.

We will now look near the line $\overline{a} + u = 0$ and will link the study of the small limit cycles (based on calculating Abelian integrals) to the study of the canard phenomenon, that will be done by means of the C^∞-foliations constructed in chapter 3.

4.2 Relation between the Abelian integrals and the center manifolds

Let us recall that for $h = 0$, we can not interpret (149) to be the principal part of (148). We will however interpret it in terms of the self-intersection of the center manifolds introduced in the previous chapter. As all these center manifolds have a mutual flat contact we can restrict to a single one; we take $\mathcal{C}_{II}(A)$, introduced as reference one in 3.3.1 and 3.3.2. We will also suppose that the transverse section T, introduced in 3.3.1, is parametrized by the variables (u, h) as defined in 3.3.1 with the extra property that h represents the value of the Hamiltonian H of (140) along the segment $[\alpha, \beta]$ (see 4.1).

We recall that $\{h = F_{II}(u, A)\}$ and $\{h = B_{II}(u, A)\}$ represent the intersection of T with respectively the forward and backward branches of $\mathcal{C}_{II}(A)$ (i.e. in positive and negative time, see figure 28).

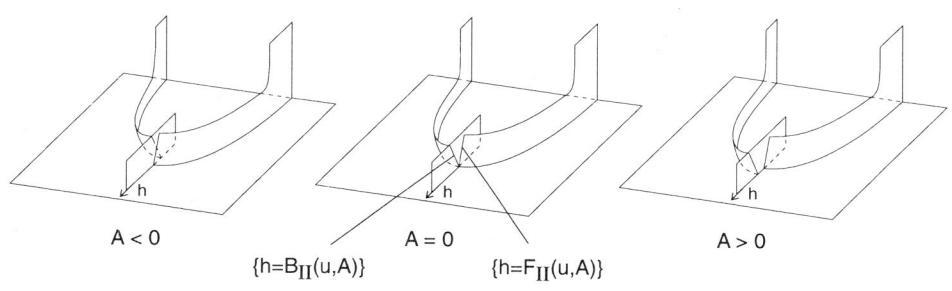

$A < 0$ $\qquad$ $A = 0$ $\qquad$ $A > 0$

$\{h=B_{II}(u,A)\}$ $\qquad$ $\{h=F_{II}(u,A)\}$

Figure 28

Pictures of $\mathcal{C}_{II}(A)$ for changing A.

Theorem 19

$$F_{II}(u, A) - B_{II}(u, A) = I_1(0)[A + u + o(|A, u|)] \tag{151}$$

Proof : As $F_{II} - B_{II}$ is C^∞ and as we already know from (150) that (151) is true for $u = 0$, it suffices to prove the statement for $A = 0$. If we write $F_{II}(u) = F_{II}(u, 0)$ and $B_{II}(u) = B_{II}(u, 0)$, we need to show that

$$F_{II}(u) - B_{II}(u) = u I_2(0) + O(u) \tag{152}$$

Let us first come back to the family $\overline{X}_1$ for $A = \overline{a} = 0$, on some large disc D. The development

$$P_{(u,0)}(h) - h = u I_2(h) + o(u) \tag{153}$$

for $h < 0$ can be interpreted geometrically as follows. If $R = \{h = h_0 + ru\} \subset T$ is a straight line, transverse to D at the point $\Gamma_{h_0} \cap T$, then the image $R' = DP_{(0,0)}(R)$ is also cutting D at the same point and is given by

$$R' = \{h = h_0 + ru + u I_2(h)\} \tag{154}$$

To prove (152) we will show the existence of a continuous line field (R_h), transverse to D, defined for $\{h \leq 0\}$ with R_0 tangent at Γ to the curve $\{h = B_{II}(u)\}$ and such that the images $R'_h = dP_{(0,0)}(R_h)$, defined for $h < 0$, extend to a continuous line field $\{R'_h \mid h \leq 0\}$ with R'_0 tangent at Γ to the curve $\{h = F_{II}(u)\}$.

Before proving the existence of (R_h), let us show that it leads to statement (152). We already know that

$$R'_h(u) - R_h(u) = u I_2(h) \tag{155}$$

for $h < 0$. By continuity at $h = 0$ this gives

$$R'_0(u) - R_0(u) = u I_2(0) \tag{156}$$

implying (152) because of the requirements on R_h.

Let us now construct the line field R_h. We choose a segment $t_0 \subset D$, transversally cutting $]s_2, s_3[\subset \sigma_1$, and parametrize it by $h \in [h_0, 0]$ for some $-1 < h_0 < 0$;

$h = H(\Gamma_h \cap t_0)$ for $h \in [h_0, 0[$.

Along t_0 we choose a line field L_h, transverse to D and with L_0 contained in $\hat{F}_0$. In the chart PR1 with coordinates (x, v, u), as in (28), we take $t_0 = \{x = 0, u = 0\}$, while the L_h are straight lines parallel to the u-axis. The L_h are transverse to the leaves $\{uv = C\}$ for $C > 0$.

We define R_h by saturating L_h, by means of the flow of X, in negative time till T. If we do the same in positive time we find R'_h. We will show that R_h is continuous and tangent to $\mathcal{C}_{II}(0)$ for $h = 0$. Similar arguments could show the same for R'_h. There is clearly no problem for $h < 0$, so we focus on $h = 0$.

The only difficulty occurs near s_2, where we can use a normal form expression (for C^1 equivalence) as given in (62) i.e.

$$z\frac{\partial}{\partial z} + v^2 f(u, v, 0)(u\frac{\partial}{\partial u} - v\frac{\partial}{\partial v}) \tag{157}$$

with f of class C^1.

Let $T_1 = \{z = z_0\}$ and $T_2 = \{v = v_0\}$ with $z_0, v_0 > 0$ sufficiently small. Take $t_1 = T_1 \cap D$, $t_2 = T_2 \cap D$ parametrized respectively by v and z near 0. The X-flow will transport L_h by saturation into a C^1 foliation inside T_1 given by

$$(z, v, u) = (z_1, \alpha(h, u), u) \tag{158}$$

As the X-flow is regular along $]s_2, s_3[$ we know that $h \mapsto \alpha(0, h)$ is a (orientation-reversing) C^1 diffeomorphism, near $h = 0$, with $\alpha(0, 0) = 0$; moreover $\alpha(0, u) = 0$ and the curves of (158) are transverse to $\{uv = C\}$ for $C > 0$.

We now saturate the foliation, given by (158), in negative time and look at the intersection with T_2. As we are only interested in tangents to these intersections at $\{u = 0\}$, we can as well look at iterates of the lines

$$(z, v, u) = (z_1, \alpha(h) + \beta(h)u, u) \tag{159}$$

with α of class C^1, β of class C^0 and $\alpha'(0) < 0$.

We can now use the formulas given in (67) to get on $T_2 = \{v = v_0\}$, graphs of functions given by

$$z(u) = z_1 \cdot exp\left(-\left(\frac{\gamma(u,h)}{v_0}\right)^2 \int\limits_{1/\gamma(u,h)}^1 sF(u\gamma(u,h)s, \frac{v_0}{\gamma(u,h)s}, 0)ds\right), \tag{160}$$

where $\gamma(u,h)$ is sure that $\overline{u} = u\gamma(u,h)$ is the solution of $\overline{u}(\alpha(h) + \beta(h)\overline{u}) = uv_0$, and hence

$$\gamma(0,h) = \frac{v_0}{\alpha(h)}, \qquad \gamma'(0,h) = -\frac{\beta(h) \cdot v_0^2}{(\alpha(h))^3} \tag{161}$$

For F we refer to (63) and we will write $F(\alpha, \beta, 0)$ when we will need derivatives of F. We are interested in the derivative of (140) with respect to u at $u = 0$, and this for $h < 0$.

Writing γ for $\gamma(u,h)$ and γ' for $\frac{\partial\gamma}{\partial u}(u,h)$, we see that this derivative is equal to expression (160) multiplied by

$$-\frac{2\gamma\gamma'}{v_0^2} \int\limits_{1/\gamma}^1 sF(u\gamma s, \frac{v_0}{\gamma s}, 0)ds - \frac{\gamma'}{\gamma v_0^2}F(u, v_0, 0)$$

$$\tag{162}$$

$$-\frac{\gamma^2}{v_0^2} \int\limits_{1/\gamma}^1 \left[s^2(\gamma + u\gamma')\frac{\partial F}{\partial \alpha}(u\gamma s, \frac{v_0}{\gamma s}, 0) - v_0\frac{\gamma'}{\gamma^2}\frac{\partial F}{\partial \beta}(u\gamma s, \frac{v_0}{\gamma s}, 0)\right] ds$$

For $u = 0$ this gives a product of :

$$z_1 \cdot exp\left(-\left(\frac{1}{\alpha(h)}\right)^2 \int\limits_{\alpha(h)/v_0}^1 sF(0, \frac{\alpha(h)}{s}, 0)ds\right) \tag{163}$$

and

$$\frac{2v_0\beta(h)}{(\alpha(h))^4} \int\limits_{\alpha(h)/v_0}^1 sF(0, \frac{\alpha(h)}{s}, 0)ds + \frac{\beta(h)}{(\alpha(h))^2}\frac{1}{v_0}F(0, v_0, 0)$$

$$\tag{164}$$

$$-\frac{1}{(\alpha(h))^2} \int\limits_{\alpha(h)/v_0}^1 \left[s^2 \cdot \frac{v_0}{\alpha(h)} \cdot \frac{\partial F}{\partial \alpha}(0, \frac{\alpha(h)}{s}, 0) + \frac{v_0\beta(h)}{\alpha(h)}\frac{\partial F}{\partial \beta}(0, \frac{\alpha(h)}{s}, 0)\right] ds$$

This product of (163) and (164) clearly tends to zero for $h \to 0$.

Hence the field of tangents to expression (160) at $u = 0$, extends in a continuous way to $\{(v, z) = (v_0 = 0)\}$, being the tangent of $\mathcal{C}_{II}(0) \cap T_2$.

Transporting this line field to T, along the flow of $-X$, corresponds to applying the derivative of the Poincaré map of $-\overline{X}_1$ from T_2 to T. Being a C^∞ map the results easily follow, including the statement on R_0, as the flow of X keeps the center manifolds $\mathcal{C}_{II}(0)$ invariant. $\qquad \square$

4.3 Explanation of the canard phenomenon by means of center manifolds

4.3.1 Canard limit periodic sets of type I

In 3.3 we have constructed the foliations of type I, II and III. Let us start by considering the $\mathcal{C}_I(y, A)$, with $y \in]0, 1/6[$ as in 3.3.1. Such a center manifold $\mathcal{C}_I(y, A)$ cuts ∂P precisely along the blowing up $\hat{\Gamma}_y^I$ of Γ_y^I, the "canard" l.p.s. of type I and height y. For the intersection $\{h = F_I(u, y, A)\}$ and $\{h = B_I(u, y, A)\}$ of $\mathcal{C}_I(y, A)$ with T, as defined in 3.3.1, we know that each solution of

$$\Delta(u, y, A) = F_I(u, y, A) - B_I(u, y, A) = 0 \tag{165}$$

corresponds to a periodic orbit, which in the blown up space M is contained in the leaf parametrized by (u, A). Such a leaf corresponds to the parameter value $(\varepsilon, a) = (u^2, uA)$.

Using (81), (82) and (101), and writing

$$\Delta_0(u, A) = F_{II}(u, A) - B_{II}(u, A), \tag{166}$$

we get

$$\Delta(u, y, A) = \Delta_0(u, A) + \Phi(u, y, A) \tag{167}$$

with

$$\Phi(u, y, A) = exp\left(-\frac{k_B(y)}{u^2}(1 + \varphi_B(u, y, A))\right) - exp\left(-\frac{k_F(y)}{u^2}(1 + \varphi_F(u, y, A))\right)$$

For the properties of these functions we refer to proposition 13 and theorem 15. This already implies that

$$\Phi(u, y, A) = exp\left(-\frac{k_B(y)}{u^2}(1 + \psi_B(u, y, A))\right) \tag{168}$$

with ψ_B having similar properties as φ_B.

Since $\Delta(u, y, A)$ and $\Delta_0(u, A)$ are both C^∞, Φ has to be C^∞ too; moreover Φ and its first derivatives are flat at $u = 0$, uniformly in (y, A) on compacta in $]0, 1/6[\times [-A_0, A_0]$ for small enough A_0.

From theorem 19 we know that $\dfrac{\partial \Delta_0}{\partial A}(0, 0)$ and $\dfrac{\partial \Delta_0}{\partial u}(0, 0)$ are different from zero, so that the implicit function theorem shows that the solution of (165) is given by a C^∞ function

$$A = A_I(u, y) \tag{169}$$

at least if we take A_0 as well as $u_I(y)$ sufficiently small in the definition of $\mathcal{D}_I$ (see (80)). $A_I(u, y) = O(u)$, uniformly in y on compacta in $]0, 1/6[$, and $A_I(u, y) < 0$ for $u > 0$.

If $A^0(u)$ denotes the solution of $\Delta_0(u, A) = 0$, then we get

$$A_I(u, y) = A^0(u) + \Psi(u, y) \tag{170}$$

with Ψ of class C^∞.

We can give a more precise expression to Ψ by combining (165) and (167) :

$$\Delta_0(u, A_0(u) + \Psi(u, y)) + \Phi(u, y, A_I(u, y)) = 0$$

together with

$$\Delta_0(u, A_0(u) + \Psi(u, y)) = \Psi(u, y) \cdot \int_0^1 \frac{\partial \Delta_0}{\partial A}(u, A_0(u) + s\Psi(u, y)) ds$$

On the domain of study we can suppose that $\dfrac{\partial \Delta_0}{\partial A} \neq 0$, so that we get

$$\Psi(u, y) = -\Phi(u, y, A_I(u, y)) \left[\int_0^1 \frac{\partial \Delta_0}{\partial A}(u, A_0(u) + s\Psi(u, y)) ds \right]^{-1} \tag{171}$$

For $A = A_I(u, y)$, the vector field in the leaf corresponding to (u, A) has a periodic orbit (a limit cycle) close to $\hat{\Gamma}_y^I$.

Let us again consider the section $\mathcal{G}_I$ from 3.3.1, transverse to $\hat{\Gamma}_y^I$ at the "horizontal" part (see figure 29) and denote by $P(v, y, A)$ the related Poincaré map for X; $F = F_I$ is the transition map from C to T in positive time, $B = B_I$ the transition map from C to T in negative time (see figure 29).

$$P = B^{-1} \circ F \tag{172}$$

We can see B^{-1} as an inverse with respect to y, taking (v, A) as parameters. We will use the fact that $\{uv = C\}$ is an invariant foliation, and we parametrize the leaves by u instead of v.

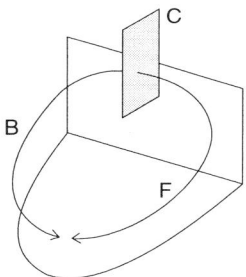

Figure 29

Study near a limit periodic set $\hat{\Gamma}_y^I$.

Let us derive (172) with respect to y, for $0 < y < 1/6$ and along the limit cycle at $A = A_I$:

$$\frac{\partial P}{\partial y}(u, y, A_I(u, y)) = \frac{\dfrac{\partial F}{\partial y}(u, y, A_I(u, y))}{\dfrac{\partial B}{\partial y}(u, y, A_I(u, y))} \tag{173}$$

and hence

$$\frac{\partial B}{\partial y}(u, y, A_I(u, y))(1 - \frac{\partial P}{\partial y}(u, y, A_I(u, y)))$$

$$= \frac{\partial B}{\partial y}(u, y, A_I(u, y)) - \frac{\partial F}{\partial y}(u, y, A_I(u, y)) = -\frac{\partial \Phi}{\partial y}(u, y, A_I(u, y)) \tag{174}$$

On the other hand

$$\frac{\partial P}{\partial y}(u, y, A_I(u, y)) = exp\left(\int_0^{T(u,y)} div\ X(\gamma_{u,y}(t))dt\right) \tag{175}$$

where $\gamma_{u,y}$ represents the limit cycle and $T(u, y)$ its period. We can of course make the calculation in the initial system $X_{\varepsilon,a}$ with $\varepsilon = u^2$, $a = uA$, and with $div\ X_{\varepsilon,a} = -(x + x^2)$. We write $dt = \dfrac{dy'}{\varepsilon(a - x)}$ (using y' instead of y in order not to confuse with the parameter y) and get

$$\int_0^{T(u,y)} div\ X(\gamma_{u,y}(t))dt = -\frac{1}{u^2}\int_{\gamma_{u,y}} \frac{(x+x^2)}{uA_I(u,y)-x}dy' \tag{176}$$

For $u \to 0$, $\gamma_{u,y}$ tends in M to the l.p.s. $\hat{\Gamma}_y^I$ (and in $I\!\!R^2$ to Γ_y^I). Let us denote

$$I(u,y) = \int_{\gamma_{u,y}} \frac{(x+x^2)}{uA_I(u,y)-x}dy' \tag{177}$$

Lemma 20 :

$I(u,y)$ tends to $k_F(y) - k_B(y)$ for $u \to 0$, uniformly in y on compacta in $]0, 1/6[$.

Proof : Starting from (173) and using (81) and (82) we find that $\dfrac{\partial P}{\partial y}(u,y,A_I(u,y))$ is the quotient of

$$\frac{1}{u^2}\left[-\frac{\partial k_F}{\partial y}(y)(1+\varphi_F) - k_F(y)\frac{\partial \varphi_F}{\partial y}\right]exp\left(-\frac{k_F(y)}{u^2}(1+\varphi_F)\right)$$

and a similar expression for $\partial B/\partial y$.

Because of proposition 13 and theorem 15, and uniformly in y on compacta in $]0, 1/6[$, we can write this quotient as

$$\frac{\dfrac{\partial k_F}{\partial y}(y)}{\dfrac{\partial k_B}{\partial y}(y)}(1+O(u))exp\left(-\frac{1}{u^2}(k_F(y) - k_B(y) + O(u))\right).$$

or even as

$$exp\left(-\frac{1}{u^2}(k_F(y) - k_B(y) + O(u))\right)$$

We of course used the fact that $A_I(u,y) = O(u)$ uniformly in y on compacta in $]0, 1/6[$. $\qquad\square$

Let us now introduce (cfr. (100)) the function

$$I(y) = -\int_{\Gamma_y^I}(1+x)dy' = \int_{x_0(y)}^{x_1(y)} x(1+x)^2dx \tag{178}$$

Because of theorem 15 we know that $k_F(y) - k_B(y) = I(y)$.
Let us recall from (168) that

$$\Phi(u, y, A_I(u, y)) = exp\left(-\frac{k_B(y)}{u^2}(1 + O(u))\right)$$

uniformly in y on compacta in $]0, 1/6[$, at least on

$$\mathcal{U}_I = \{(u, y) \mid 0 \le y \le 1/6, 0 \le u \le u_I(y)\} \tag{179}$$

with $u_I(y)$ like in (80) and sufficiently small.

In the next theorem we let $\Gamma_{u,A}$ represent the unique limit cycle for the choice of parameters (u, A); it is equal to $\Gamma_{\varepsilon,a}$ with $(\varepsilon, a) = (u^2, uA)$, as defined in chapter 1. We will also use the mapping Γ sending $(u, y) \in \mathcal{U}_I$ to the limit cycle $\Gamma_{u,A_I(u,y)}$ for $u \ne 0$ and to Γ_y^I for $u = 0$.

Theorem 21

For all $(u, y) \in \mathcal{U}_I \cap \{u > 0\}$, $\dfrac{\partial A_I}{\partial y}(u, y) < 0$, and the curves $\{(u, A_I(u, y))\}$ define a C^∞ foliation $\mathcal{F}_I$ of the region

$$\mathcal{V}_I = \{(u, A) \mid A = A_I(u, y), (u, y) \in \mathcal{U}_I\} \tag{180}$$

The mapping $Y : \mathcal{V}_I \to \mathbb{R}$, given by $Y(u, A_I(u, y)) = y$ defines a fibration on $]0, 1/6[$, C^∞ outside $(0, 0)$.
If $c : u \mapsto (u, A(u))$ defines a continuous path in $\mathcal{V}_I$ with $A(0) = 0$, then

$$\Gamma_{u,A(u)} \to \Gamma_y^I \quad iff \quad Y(u, A(u)) \to y \tag{181}$$

Proof

As $A_I(u, y)$ is defined implicitly by (165) we get

$$\frac{\partial \Delta}{\partial A}(u, y, A_I(u, y)) \cdot \frac{\partial A_I}{\partial y}(u, y) + \frac{\partial \Delta}{\partial y}(u, y, A_I(u, y)) = 0$$

and hence (see (167) :

$$\frac{\partial \Delta}{\partial A}(u, y, A_I(u, y)) \cdot \frac{\partial A_I}{\partial y}(u, y) = -\frac{\partial \Phi}{\partial y}(u, y, A_I(u, y)) \tag{182}$$

Taking $u_I(y)$ sufficiently small, we have that $\dfrac{\partial \Delta}{\partial A} \neq 0$ and hence it has the sign of $I_1(0)$ - (see (151)) - which is negative. From (174), (178) and proposition 13, we get that the right-hand side of (182) is strictly positive for $u \neq 0$, implying that $\dfrac{\partial A_I}{\partial y}(u,y) < 0$ for $u \neq 0$.

From this inequality immediately follows the statement on the foliation $\mathcal{F}_I$ and on the related fibration given by Y. Moreover $(u,y) \mapsto (u, A_I(u,y))$ is a homeomorphism from $\mathcal{U}_I \cap \{u > 0\}$ onto its image. Now the mapping Γ, as defined above, is continuous for the Hausdorff metric on compact subsets of $\mathbb{R}^2$. Indeed a limit cycle $\Gamma_{u,A_I(u,y)}$ is defined to be the self-intersection of the manifold $\mathcal{C}_I(y, A_I(u,y))$. As this center manifold depends in a C^∞ way on (y,A) and as A_I itself is C^∞ it is clear that $\Gamma_{u,A_I(u,y)}$ depends continuously on (u,y) for $u \neq 0$.

The continuity at $(0,y)$, for some $y \in]0, 1/6[$ follows from the following : Γ_y^I depends continuously on y, while the convergence of $\Gamma_{u,A_I(u,y)}$ to Γ_y^I, for $u \to 0$, is uniform in y near each value $y = y_0$. Indeed, $\mathcal{C}_I(y,A)$ depends continuously on (y,A) and within such a manifold, the distance of whatsoever orbit to the boundary $\mathcal{C}_I(y,A) \cap \partial P$, tends to 0, uniformly in y, near the value $y = y_0$, when the initial condition has a coordinate $u \to 0$.

It is now clear that (181) follows from the continuity of Γ. If along a path c, $Y(u, A(u)) \to y$, then $\Gamma_{u,A(u)} \to \Gamma_y^I$.

Conversely, if $\Gamma_{u,A(u)} \to \Gamma_y^I$, and $Y(u, A(u))$ would not converge to y, then either we could find a subsequence $(u_n)_n \to 0$ with $Y(u_n) \to y' \in]0, 1/6[\setminus \{y\}$ or $Y(u, A(u))$ would converge to 0 or to $1/6$. In the first case, by continuity of Γ, we would obtain $\Gamma_{u_n,A(u_n)} \to \Gamma_{y'}^I$, contradicting $\Gamma_{u,A(u)} \to \Gamma_y^I$; in the second case the result is obvious.

$\square$

Condition (181) is not intrinsic since it uses the foliation $\mathcal{F}_I$, which is not defined in a unique way. We can however change condition (181) by an intrinsic one describing the order of flatness of c with respect to the reference curve $A = A^\circ(u)$.

Theorem 22

With the notations of theorem 21, let $c : u \mapsto (u, A(u))$ be a continuous path in $\mathcal{V}_I$

with $A(0) = 0$ and $A^\circ(u) < A(u) < 0$ for $u > 0$, then

$$\Gamma_{u, A(u)} \to \Gamma_y^I \qquad iff \qquad \lim_{u \to 0}(-u^2 \log(A(u) - A^\circ(u))) = k_B(y) \tag{183}$$

Proof

Let us first recall from (170) that $A_I(u, y) = A^\circ(u) + \Psi(u, y)$; from (171) and (168) follows that

$$\Psi(u, y) = exp\left(-\frac{k_B(y)}{u^2}(1 + O(u))\right) \tag{184}$$

and also that $\dfrac{\partial \Phi}{\partial y}(u, y) < 0$ on $\mathcal{U}_I \cap \{u > 0\}$ for $u_I(y)$ sufficiently small.

Let now $\Gamma_{u, A(u)} \to \Gamma_y^I$, which - by (181) - is equivalent top $Y(u, A(u)) \to y$. If we consider $y_1 < y < y_2$ then, for sufficiently small u :

$$A^\circ(u) + \Psi(u, y_2) \le A(u) \le A^\circ(u) + \Psi(u, y_1) \tag{185}$$

and hence

$$k_B(y_1) \le -u^2 \log(A(u) - A^\circ(u)) \le k_B(y_2)$$

From this easily follows the "only if" part in (183).

Conversely, if $-u^2 \log(A(u) - A^\circ(u)) \to k_B(y)$ and if $y_1 < y < y_2$ are given, then for sufficiently small u, we have (185) and hence $y_1 \le Y(u) < y_2$ for the same u. This will induce the "if" part of (183). $\qquad \square$

It is clear that theorem 22 is a restatement of part 2) in theorem 1, if we use $(\varepsilon, a) = (u^2, uA)$, and define $k(y) = k_B(y)$ as we did in (106).

This finishes the study of the canard cycles of type I on a domain $\mathcal{V}_I$ in (u, A)-space (see (180) and figure 30).

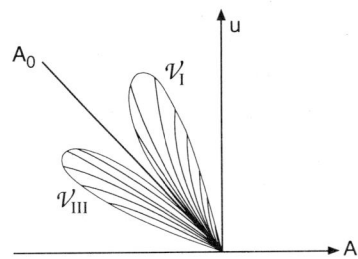

Figure 30
Regions $\mathcal{V}_I$ and $\mathcal{V}_{III}$ in the (u, A)-plane.

4.3.2   Canard limit periodic sets of type III

A similar treatment as in 4.3.1 can be done to study the limit cycles near the limit periodic sets $\Gamma_{y.}^{III}$, by looking at

$$\tilde{\Delta}(u, y, A) = F_{III}(u, y, A) - B_{III}(u, y, A) \tag{186}$$

and writing

$$\tilde{\Delta}(u, y, A) = \Delta_0(u, A) + \tilde{\Phi}(u, y, A) \tag{187}$$

this time using the expressions (136) and (137) for F_{III} and B_{III}.

Again it will be possible to find a parametrization

$$A_{III}(u, y) = A^\circ(u) - \tilde{\Psi}(u, y) \tag{188}$$

of some region $\mathcal{V}_{III}$ (like in figure 30) situated to the left of $\{(u, A^\circ(u)\}$, with $\tilde{\Psi}$ of class C^∞ and defined on

$$\mathcal{U}_{III} = \{(u, y) \mid 0 \leq y \leq 1/6, 0 \leq u \leq u_{III}(y)\} \tag{189}$$

with u_{III} like in (135) and sufficiently small. Like in (184) and on $\mathcal{U}_{III}$ we have

$$\tilde{\Psi}(u, y) = exp\left(-\frac{k_B(y)}{u^2}(1 + O(u))\right) \tag{190}$$

as well as $\dfrac{\partial \tilde{\Psi}}{\partial y}(u, y) < 0$, with exactly the same function $k_B(y)$ as in (184) and $O(u)$ having similar properties as in (184).

The function $A_{III}(u, y)$ has the property that

$$\Gamma_{u, A_{III}(u,y)} \to \Gamma_y^{III} \tag{191}$$

when $u \to 0$.

All proofs are similar to the ones given for $A_I(u, y)$ except that we have to rely on proposition 17 instead of proposition 13.

We also obtain similar statements as the ones in theorem 21 and 22, namely that in

$$\mathcal{V}_{III} = \{(u, A) \mid A = A_{III}(u, y), (u, y) \in \mathcal{U}_{III}\} \tag{192}$$

the curves $\{(u, A_{III}(u, y)\}$ form a C^∞ foliation $\mathcal{F}_{III}$, while $Y : \mathcal{V}_{III} \to I\!\!R$, given by $Y(u, A_{III}(u, y)) = y$, defines a fibration on $]0, 1/6[$, C^∞ outside $(0, 0)$. See figure 30 for a representation of $\mathcal{V}_{III}$. The mapping $\overline{A}_{III} : \mathcal{U}_{III} \to \mathcal{V}_{III}$, defined by $(u, y) \mapsto (u, A_{III}(u, y))$, is a homeomorphism (for $u > 0$). If $c : u \to (u, A(u))$ defines a continuous path in $\mathcal{V}_{III}$ with $A(0) = 0$ and $A(u) < A^\circ(u)$ for $u > 0$, then

$$\Gamma_{u, A(u)} \to \Gamma_y^{III} \quad \text{iff} \quad Y(u, A(u)) \to y \tag{193}$$

which can only happen iff

$$\lim_{u \to 0}(-u^2 \log(A^\circ(u) - A(u))) = k_B(y)$$

4.3.3 Canard limit periodic sets of type II

Here the situation is somewhat different since we only have one l.p.s. of type II and after desingularization it corresponds to a family of l.p.s.
We consider the $\mathcal{C}_{II}(\theta, A)$ with $\theta \in]0, \pi[$ as in 3.3.2. For the intersections

$\{h = F_{II}(u, \theta, A)\}$ and $\{h = B_{II}(u, \theta, A)\}$ of $\mathcal{C}_{II}(\theta, A)$ with T, as defined in 3.3.2, we know that each solution of

$$\Delta(u, \theta, A) = F_{II}(u, \theta, A) - B_{II}(u, \theta, A) = 0 \tag{194}$$

corresponds to a periodic orbit $\Gamma_{u,A}$ in the leaf parametrized by (u, A). We again get

$$\Delta(u, \theta, A) = \Delta_0(u, A) + \Phi(u, \theta, A) \tag{195}$$

using the expression (120) and (121) and leading to a parametrization

$$A_{II}(u, \theta) = A^\circ(u) + \Psi(u, \theta) \tag{196}$$

of some region $\mathcal{V}_{II}$ (see figure 33) covering $\{(u, A^\circ(u))\}$, for u small, with Ψ of class C^∞ and defined on

$$\mathcal{U}_{II} = \{(u, \theta) \mid \theta < \theta < \pi, 0 \leq u \leq u_{II}\} \tag{197}$$

with u_{II} a number like in (117) and sufficiently small.

Again one can prove that $\overline{A}_{II} : \mathcal{U}_{II} \to \mathcal{V}_{II}$, $(u, \theta) \mapsto (u, A_{II}(u, \theta))$, is a homeomorphism (for $u > 0$) and the curves $\{(u, A_{II}(u, \theta))\}$ define a C^∞ foliation $\mathcal{F}_{II}$ of $\mathcal{V}_{II}$. The mapping $\Theta : \mathcal{V}_{II} \to \mathbb{R}$, given by $\Theta(u, A_{II}(u, \theta)) = \theta$, defines a fibration on $]0, \pi[$, C^∞ outside $(0, 0)$.

Again, if $c : u \mapsto (u, A(u))$ defines a continuous path in $\mathcal{V}_{II}$ with $A(0) = 0$, then

$$\Gamma_{u, A(u)} \to \hat{\Gamma}_\theta^{II} \Leftrightarrow \Theta(u, A(u)) \to \theta \tag{198}$$

and $\hat{\Gamma}_\theta^{II}$ stands for the blowing up of Γ^{II}, corresponding to the parameter θ.

However the estimates (120) and (121) in proposition 16 do not permit to characterize this in a precise way like we did in (183) for the Γ_y^I, but only permit to show that

$$\Gamma_{u, A(u)} \to \Gamma^{II} \Leftrightarrow \underline{\lim}(-u^2 \mid A^\circ(u) - A(u) \mid) \geq k_B(1/6) \tag{199}$$

4.3.4   Bringing the foliations together (as a final step)

In 4.3.1 - 4.3.3 we obtained foliations $\mathcal{F}_I$ in $\mathcal{V}_I$, $\mathcal{F}_{II}$ in $\mathcal{V}_{II}$ and $\mathcal{F}_{III}$ in $\mathcal{V}_{III}$, constructed by means of mappings A_I, A_{II} and A_{III} defined on respectively $\mathcal{U}_I$, $\mathcal{U}_{II}$, $\mathcal{U}_{III}$ (see (179), (197) and (189)) with $u_I(y)$, u_{II} and $u_{III}(y)$ sufficiently small.

On the other hand we have, within P, sections $\mathcal{G}_I$, $\mathcal{G}_{II}$ and $\mathcal{G}_{III}$ foliated respectively by segments γ_y (see 3.3.1 and figure 22), sections γ_θ (see 3.3.2 and figure 24) and sections $\tilde{\gamma}_y$ (see 3.3.3 and figure 25), and also parametrized by the same $\mathcal{U}_I$, $\mathcal{U}_{II}$ and $\mathcal{U}_{III}$ (let us say by means of a parametrization P_I, P_{II}, P_{III}). For $i = I, II, III$, we can hence see $\mathcal{F}_i$ as the image of the foliation by segments in $\mathcal{G}_i$ using a mapping $\tilde{A}_i : \mathcal{G}_i \to \mathcal{V}_i$, where $\tilde{A}_i = \overline{A}_i \circ (P_i)^{-1}$ and $\overline{A}_i(u, z) = (u, A_i(u, z))$ with $z = \theta$ or y, at least for $u > 0$.

The fact that we take $u_I(y)$, u_{II} and $u_{III}(y)$ small means that at certain steps of the construction we diminish the $\mathcal{G}_i$ by making the respective segments γ_y, γ_θ and $\tilde{\gamma}_y$ shorter.

There is still an amount of freedom in the construction permitting to take $\mathcal{G}_{II}$ in a way that its boundary $\partial \mathcal{G}_{II}$ contains one of the segments $\gamma_y \subset \mathcal{G}_I$ and one of the segments $\tilde{\gamma}_y \subset \mathcal{G}_{III}$ (see figure 31); moreover in $\mathcal{G}_I \cap \mathcal{G}_{II}$ and in $\mathcal{G}_{III} \cap \mathcal{G}_{II}$ we can take the segments γ_θ to cut transversally the respective segments γ_y and $\tilde{\gamma}_y$.

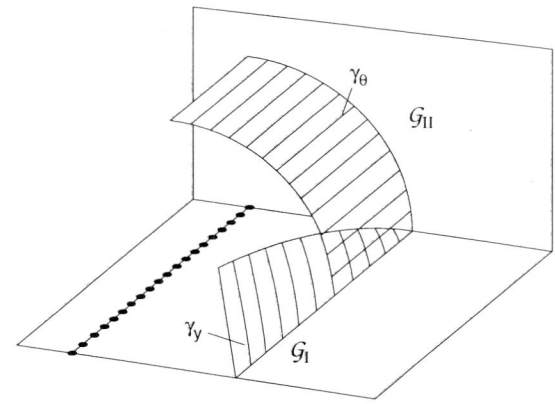

Figure 31
Link of $\mathcal{G}_I$ and $\mathcal{G}_{II}$.

A similar construction can be done with respect to the passage from the l.p.s. Γ_y^I to the "small l.p.s.".

We consider a rectangle $\mathcal{G}_S \subset P$, as in figure 32, with the property that one side is a straight line segment L in D going from the point β in σ_1 (see 3.3.1 and figure 22) to $(0,0)$ in D and being transverse to the closed orbits of $\overline{X}_1 \mid \overline{D}_{(1,0)}$ (see (138)).

Along this side we can use the value of the Hamiltonian H (see (140)) as a parameter; $h \in [-1, 0]$ with $h = -1$ representing the singularity and $h = 0$ representing σ_1 as well as the connection Γ.

We take one side of $\mathcal{G}_S$ to be $\{(\overline{x}, \overline{y}, u) = (0, 0, u)\}$ and another one inside $\hat{F}_0$. We parametrize G_S by means of (u, h), restrict (u, h) to

$$\mathcal{U}_S = \{(u, h) \mid 0 \leq u \leq u_S, -1 \leq h \leq 0\} \tag{200}$$

and take $u_S > 0$ sufficiently small, where needed.

Let $P_S : \mathcal{U}_S \to \mathcal{G}_S$ denote the parametrization.

We foliate, in a C^∞ way, $\mathcal{G}_S$ by means of line segments γ_h, having $(0, h)$ as base point.

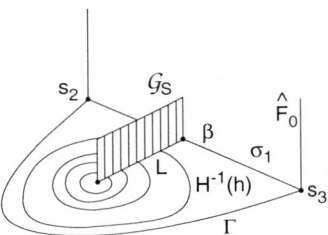

Figure 32

$\mathcal{G}_S$, *foliated by segments* γ_h.

We can arrange the construction in a way that the fourth side of $\mathcal{G}_S$ contains one of the segments $\gamma_y \subset \mathcal{G}_I$.

As usual, an extra dimension can be added by considering

$$\mathcal{D}_S = \{(u, h, A) \mid 0 \leq u \leq u_S, -1 < h < 0, -A_0 \leq A \leq A_0\} \tag{201}$$

The saturation of the segments γ_h in forward and backward time, analogously to the definition of F_i and B_i with $i = I, II, III$, leads to the definition of $\{h' = F_S(u, h, A)\}$ and $\{h' = B_S(u, h, A)\}$ on the same section T. (We use h' instead of h as variable on T in order not to confuse with the parameter h). These functions are C^∞ outside

$\{h = 0\}$ and extend in a C^1 way to 0 at $\{h = 0\}$, as a consequence of the theorems 18 and 19. Theorem 18 implies that the equation

$$F_S(u, h, A) - B_S(u, h, A) = 0 \tag{202}$$

defines a C^∞ function $A_S : \mathcal{U}_S \cap \{h < 0\} \to I\!\!R$ with

$$A_S(u, h) = -\frac{I_2(h)}{I_1(h)} u + o(u) \tag{203}$$

representing the solution of (202). By theorem 19, A_S extends in a C^1 way at $\{h = 0\}$. Let us recall from theorem 18 that

$$A_S(u, 0) = 0, \tag{204}$$

representing the line of Hopf bifurcations, and that $I_2(h)/I_1(h)$ increases in a monotone way from 0 to 1 when h increases from -1 to 0.
Let us define

$$\mathcal{V}_S = \{(u, A) \mid A = A_S(u, h), (u, h) \in \mathcal{U}_S \backslash \{h = 0, u > 0\}\} \tag{205}$$

For $u \to 0$ we clearly have

$$\hat{\Gamma}_{u, A_S(u, h)} \to \Gamma_h \tag{206}$$

for the Hausdorff metric. Again $\overline{A}_S$, defined by $\overline{A}_S(u, h) = (u, A_S(u, h))$, is a homeomorphism for $u > 0$ and $h < 0$, and $\tilde{A}_S = \overline{A}_S \circ P_S^{-1}$ is (for $u > 0$ and $h < 0$) a homeomorphism that sends the foliation $\{\gamma_h\}$ to the foliation defined by the curves $\{(u, A_S(u, h))\}$.
If we define $\tilde{H} : \mathcal{V}_S \to I\!\!R$ to be $\tilde{H}(u, A_S(u, h)) = h$, and $c : u \to (u, A(u))$ defines a continuous path in $\mathcal{V}_S$, with $A(0) = 0$, then

$$\hat{\Gamma}_{u, A(u)} \to \Gamma_h \qquad \text{iff} \qquad \tilde{H}(u, A(u)) \to h \tag{207}$$

This is only possible iff

$$A(u)/u \to -I_2(h)/I_1(h) \tag{208}$$

In figure 33 we represent the different foliations in $(u, \overline{A})$-space for $\overline{A} = A/u$.

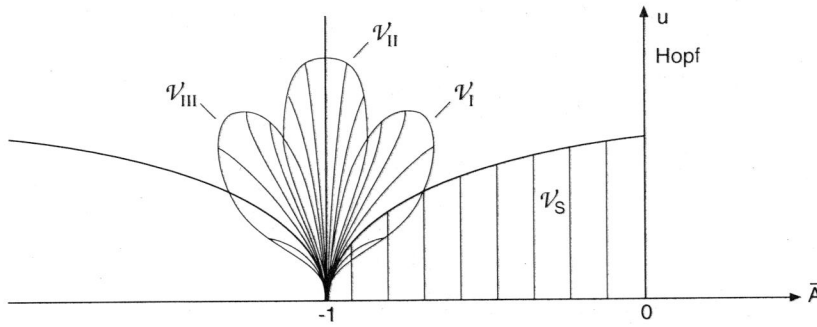

Figure 33
Different foliations in $(u, \overline{A})$-space.

The knowledge on the different foliations in $\mathcal{V}_S, \mathcal{V}_I, \mathcal{V}_{II}, \mathcal{V}_{III}$ and on the orbits through the initial conditions in $\mathcal{D}_S, \mathcal{D}_I, \mathcal{D}_{II}$ and $\mathcal{D}_{III}$, easily implies theorem 2 as well the first four statements in theorem 1 (w.r.t. $\Gamma_0, \Gamma_y^I, \Gamma^{II}$ and Γ_y^{III}). There only remains to prove the statement w.r.t. Γ_B.

The preceding results already imply that in order to have Γ_B as l.p.s. along a continuous curve c given by $\{(u, A(u))\}$ with $A(0) \in [-1/2, 0]$, we need to have $A(u) < A^\circ(u)$ for small $u > 0$, as well as $\overline{\lim}(-u^2 \log(A^\circ(u) - A(u))) \leq 0$.

The proof of the opposite follows by e.g. showing that for any $a \in [-1/2, 0[$, $\Gamma_{\varepsilon, a}$ tends to Γ_B (for the Hausdorff metric), when $\varepsilon \to 0$. This fact is a direct consequence of the geometric picture that we obtained by blowing up (see chapter 2).

With the use of center manifolds and related techniques as developed in the chapters 2 and 3, we could even make a further distinction in between the different shapes that a "big" l.p.s. Γ_B can have after desingularisation, and the way that c needs to tend to $\{\varepsilon = 0\}$ in order to get those specific possibilities. There is however no need to do this in view of proving the theorems 1 and 2.

References

[ALGM]   A. Andronov, E. Leontovich, I. Gordon, A. Maier, *Theory of bifurca-
            tions of dynamical systems on a plane*, I.P.S.T., Jerusalem, 1971.

[BCD]   E. Benoit, J.L. Callot, F. Diener, M. Diener, *Chasse au canard*, Col-
            lectanea Mathematica, 31-32 (1-3), 37-119, 1981.

[B]   R. Bogdanov, *Versal deformation of a singularity of a vector field on
            the plane in the case of zero eigenvalues*, Selecta. Math. Soviet. 1, 389-
            421, 1981.

[Bon]   P. Bonckaert, *Partially hyperbolic fixed points with constraints*, to ap-
            pear in Trans. Amer. math. Soc.

[C]   W. Coppel, *Some quadratic systems with at most one limit cycle*, in
            Dynamics Reported Vol 2, Dynam. Report. Ser. Dynam. Syst. Appl.
            2, Wiley, Chicester, 61-68, 1989.

[DeR]   Z. Denkowska, R. Roussarie, *A Method of desingularization for analytic
            two-dimensional vector field families*, Bol. Soc. Bras. Mat., Vol 22, No
            1, 93-126, 1991.

[D]   F. Dumortier, *Techniques in the theory of local bifurcations : blow-
            up, normal forms, nilpotent bifurcations, singular perturbations* (notes
            written with B. Smits), in "Bifurcations of Periodic Orbits and vector
            fields", D. Schlomiuk edit., NATO ASI-series C408, 19-73, 1993.

[DR]   F. Dumortier, R. Roussarie, *Tracking limit cycles escaping from rescal-
            ing domains*, in : Proc. Intern. Conf. Dynam. Systems and Related
            Topics, Adv. Ser. Dynam. Systems 9, World Scientific, Singapore, 80-
            99, 1991.

[E]   W. Eckhaus, *Relaxation oscillations including a standard chase on
            French ducks*, in Asymptotic Analysis II, Springer Lect. Notes in Math.
            985, 449-494, 1983.

[HPS]   M. Hirsch, C. Pugh, M. Shub, *Invariant manifolds*, Lecture Notes in
            Math. 583, Springer-Verlag, 1977.

[K]   A. Kelley, *The stable, center-stable, center, center-unstable and unstable manifolds*, Appendix C in R. Abraham, J. Robbin : Transversal mappings and flows, Benjamin, New York, 1967.

[LMP]   A. Lins, W. de Melo, C.C. Pugh, *On Liénard's equation*, Proc. Symp. Geom. and Topol., Springer Lecture Notes 597, 335-357, 1977.

[R]   R. Roussarie, *Desingularization of unfoldings of cuspidal loops*. In : Geometry and Analysis in Non-linear Dynamics, ed. H.W. Broer and F. Takens, Pitman Research Notes in Math. Series 222, 41-55, 1992.

[S]   J. Sotomayor, *Liçoes de equacóes deferenciais ordinárias*, Projeto Euclides, CNPq, 1979.

[T]   F. Takens, *Partially hyperbolic fixed points*, Topology 10, 133-147, 1971.

[V]   A. Vanderbauwhede, *Center manifolds, normal forms and elementary bifurcations*, in : Dynamic Reported Vol. 2, Dynam. Report Ser. Dynam. Syst. Appl. 2, Wiley, Chicester, 89-169, 1989.

[vG]   S. van Gils, personal communication.

APPENDIX

ON THE PROOF OF THEOREM 18

by

Chengzhi Li

(Using the method from : Chow, Li and Wang, *Uniqueness of Periodic Orbits in Some Vector Fields with cod. 2 singularities*, J.D.E., 77 n°2 (1989) - p. 231-253).

Let $X_{\lambda,\mu}$ be a family of vector fields, with equation :

$$\begin{cases} \dot{x} = (y - \dfrac{x^2}{2} + \mu x^3)e^{-y} \\[2mm] \dot{y} = (-x + \lambda)e^{-y} \end{cases}$$

If $\lambda = \mu = 0$, $X_{0,0}$ is the Hamiltonian vector field with Hamiltonian :

$$H(x,y) = -(y - \frac{x^2}{2} + 1)e^{-y}.$$

Let $\Gamma_h = \{H(x,y) = h\}$, for $h > -1$ be the Hamiltonian cycle at level h. The center $(0,0)$ corresponds to $H(0,0) = -1$.

Let

$$I_1(h) = \int_{\Gamma_h} e^{-y}dx = \int_{\Gamma_h} e^{-y}\,xdy$$

$$I_2(h) = \frac{1}{3}\int_{\Gamma_h} e^{-y}\,x^3\,dy \qquad \text{and} \qquad P(h) = \frac{I_2(h)}{I_1(h)}.$$

Then, $P(h) > 0$ for $h > -1$ and $\lim\limits_{h \to -1} P(h) = 0$.

We want to prove that $P'(h) > 0$ for $h > -1$.

Suppose the contrary. Then there exists a smallest $h_0 > -1$ such that $P'(h_0) = 0$.

We will prove that : $P(h) - P(h_0) > 0$ for $0 < | h - h_0 | << 1$ and $h < h_0$ $(*)$. This will give a contradiction.

Note that :

$$
\begin{aligned}
P(h) - P(h_0) &= \frac{I_2(h)}{I_1(h)} - \frac{I_2(h_0)}{I_1(h_0)} \\
&= \frac{\xi(h) - \xi(h_0)}{I_1(h).I_1(h_0)} \\
&= \frac{\xi'(\theta).(h - h_0)}{I_1(h)I_1(h_0)}
\end{aligned}
$$

for some θ between h_0 and h when

$$
\xi(h) = I_1(h_0) \ I_2(h) - I_2(h_0) \ I_1(h).
$$

So, we have :

$$
P(h) - P(h_0) = \frac{h - h_0}{I_1(h)} \ Q(\theta) \quad (**)
$$

where $Q(h) = I_2'(h) - P(h_0)I_1'(h)$.

It is easy to see that $P'(h_0) = 0$ implies that $Q(h_0) = 0$.

The equation :

$$
\Gamma_h = \{-(y - \frac{x^2}{2} + 1)e^{-y} = h\}
$$

for Γ_h defines two functions :
$x = x_i(y, h)$ with $x_1 < 0$, $x_2 > 0$, $x_2 = -x_1$ and $x_i' = \frac{dx_i}{dh} = \frac{e^y}{x_i}$, $i = 1, 2$. Let us write x' for both x_1' and x_2'.

One has:

$$Q(h) = I_2'(h) - P(h_0)I_1'(h) = \int_{\Gamma_h} e^{-y}\, x^2\, x'dy - P(h_0)\int_{\Gamma_h} e^{-y}x'dy$$
$$= \int_{\Gamma_h} \frac{x^2 - P(h_0)}{x} dy.$$

Let $h < h_0$ and $|\, h - h_0\,| << 1$. The curve Γ_{h_0} contains Γ_h in its interior and the lines $x = \pm\sqrt{P(h_0)}$ must intersect Γ_{h_0} since $Q(h_0) = 0$ ($P(h_0) > 0$). The annulus D between Γ_{h_0} and Γ_h is decomposed in 4 disks D_1, D_2, D_3, D_4 by these 2 lines. Observe that the parabola $\{y = \frac{x^2}{2}\}$ does not intersect disks D_2, D_4 and that the axis $\{x = 0\}$ does not intersect disks D_1, D_3 (See figure 34).

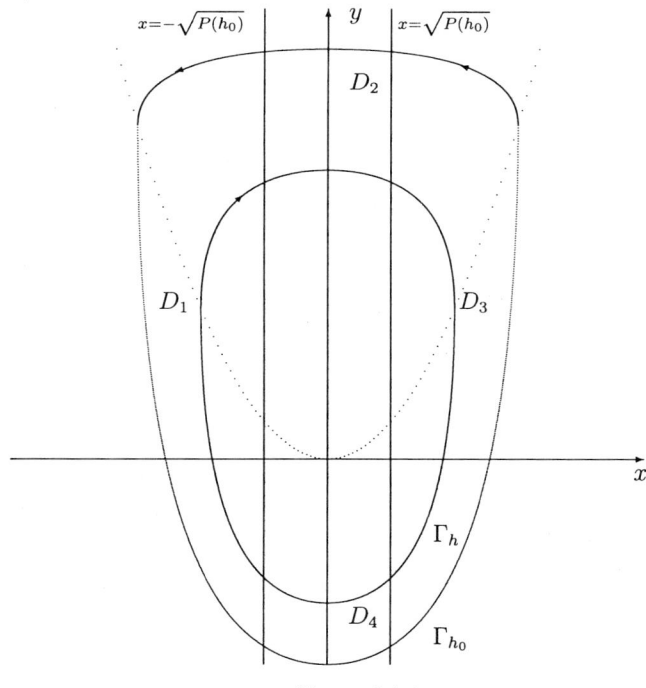

Figure 34

Now, one has :

$$
\begin{aligned}
Q(h) &= Q(h) - Q(h_0) \\
&= \int_{\Gamma_h} \frac{x^2 - P(h_0)}{x}\, dy - \int_{\Gamma_{h_0}} \frac{x^2 - P(h_0)}{x}\, dy \\
&= \int_{\partial D^+} \frac{x^2 - P(h_0)}{x}\, dy \\
&= \int_{\partial D_1^+ \cup \partial D_3^+} \frac{x^2 - P(h_0)}{x}\, dy + \int_{\partial D_2^+ \cup \partial D_4^+} \frac{x^2 - P(h_0)}{x^2/2 - y}\, dx \\
&= \int\!\!\int_{D_1 \cup D_3} \frac{x^2 + P(h_0)}{x^2}\, dx dy + \int\!\!\int_{D_2 \cup D_4} \frac{P(h_0) - x^2}{(\dfrac{x^2}{2} - y)^2}\, dx dy > 0.
\end{aligned}
$$

On the other hand, $I_1(h) < 0$ for $h < h_0$, $\mid h - h_0 \mid << 1$, hence $P(h) - P(h_0) > 0$ for $0 < \mid h - h_0 \mid << 1$, $h < h_0$. This ends the proof of assumption $(*)$.

Editorial Information

To be published in the *Memoirs*, a paper must be correct, new, nontrivial, and significant. Further, it must be well written and of interest to a substantial number of mathematicians. Piecemeal results, such as an inconclusive step toward an unproved major theorem or a minor variation on a known result, are in general not acceptable for publication. *Transactions* Editors shall solicit and encourage publication of worthy papers. Papers appearing in *Memoirs* are generally longer than those appearing in *Transactions* with which it shares an editorial committee.

As of January 31, 1996, the backlog for this journal was approximately 5 volumes. This estimate is the result of dividing the number of manuscripts for this journal in the Providence office that have not yet gone to the printer on the above date by the average number of monographs per volume over the previous twelve months, reduced by the number of issues published in four months (the time necessary for preparing an issue for the printer). (There are 6 volumes per year, each containing at least 4 numbers.)

A Copyright Transfer Agreement is required before a paper will be published in this journal. By submitting a paper to this journal, authors certify that the manuscript has not been submitted to nor is it under consideration for publication by another journal, conference proceedings, or similar publication.

Information for Authors and Editors

Memoirs are printed by photo-offset from camera copy fully prepared by the author. This means that the finished book will look exactly like the copy submitted.

The paper must contain a *descriptive title* and an *abstract* that summarizes the article in language suitable for workers in the general field (algebra, analysis, etc.). The *descriptive title* should be short, but informative; useless or vague phrases such as "some remarks about" or "concerning" should be avoided. The *abstract* should be at least one complete sentence, and at most 300 words. Included with the footnotes to the paper, there should be the 1991 *Mathematics Subject Classification* representing the primary and secondary subjects of the article. This may be followed by a list of *key words and phrases* describing the subject matter of the article and taken from it. A list of the numbers may be found in the annual index of *Mathematical Reviews*, published with the December issue starting in 1990, as well as from the electronic service e-MATH [**telnet e-MATH.ams.org** (or **telnet 130.44.1.100**). Login and password are **e-math**]. For journal abbreviations used in bibliographies, see the list of serials in the latest *Mathematical Reviews* annual index. When the manuscript is submitted, authors should supply the editor with electronic addresses if available. These will be printed after the postal address at the end of each article.

Electronically prepared papers. The AMS encourages submission of electronically prepared papers in $\mathcal{A}_{\mathcal{M}}\mathcal{S}$-TeX or $\mathcal{A}_{\mathcal{M}}\mathcal{S}$-LaTeX. The Society has prepared author packages for each AMS publication. Author packages include instructions for preparing electronic papers, the *AMS Author Handbook*, samples, and a style file that generates the particular design specifications of that publication series for both $\mathcal{A}_{\mathcal{M}}\mathcal{S}$-TeX and $\mathcal{A}_{\mathcal{M}}\mathcal{S}$-LaTeX.

Authors with FTP access may retrieve an author package from the Society's Internet node **e-MATH.ams.org** (130.44.1.100). For those without FTP access, the author package can be obtained free of charge by sending e-mail to `pub@math.ams.org` (Internet) or from the Publication Division, American Mathematical Society, P.O. Box 6248, Providence, RI 02940-6248. When requesting an author package, please specify $\mathcal{AMS}$-TEX or $\mathcal{AMS}$-LATEX, Macintosh or IBM (3.5) format, and the publication in which your paper will appear. Please be sure to include your complete mailing address.

Submission of electronic files. At the time of submission, the source file(s) should be sent to the Providence office (this includes any TEX source file, any graphics files, and the DVI or PostScript file).

Before sending the source file, be sure you have proofread your paper carefully. The files you send must be the EXACT files used to generate the proof copy that was accepted for publication. For all publications, authors are required to send a printed copy of their paper, which exactly matches the copy approved for publication, along with any graphics that will appear in the paper.

TEX files may be submitted by email, FTP, or on diskette. The DVI file(s) and PostScript files should be submitted only by FTP or on diskette unless they are encoded properly to submit through e-mail. (DVI files are binary and PostScript files tend to be very large.)

Files sent by electronic mail should be addressed to the Internet address `pub-submit@math.ams.org`. The subject line of the message should include the publication code to identify it as a Memoir. TEX source files, DVI files, and PostScript files can be transferred over the Internet by FTP to the Internet node `e-math.ams.org` (130.44.1.100).

Electronic graphics. Figures may be submitted to the AMS in an electronic format. The AMS recommends that graphics created electronically be saved in Encapsulated PostScript (EPS) format. This includes graphics originated via a graphics application as well as scanned photographs or other computer-generated images.

If the graphics package used does not support EPS output, the graphics file should be saved in one of the standard graphics formats—such as TIFF, PICT, GIF, etc.—rather than in an application-dependent format. Graphics files submitted in an application-dependent format are not likely to be used. No matter what method was used to produce the graphic, it is necessary to provide a paper copy to the AMS.

Authors using graphics packages for the creation of electronic art should also avoid the use of any lines thinner than 0.5 points in width. Many graphics packages allow the user to specify a "hairline" for a very thin line. Hairlines often look acceptable when proofed on a typical laser printer. However, when produced on a high-resolution laser imagesetter, hairlines become nearly invisible and will be lost entirely in the final printing process.

Screens should be set to values between 15% and 85%. Screens which fall outside of this range are too light or too dark to print correctly.

Any inquiries concerning a paper that has been accepted for publication should be sent directly to the Editorial Department, American Mathematical Society, P. O. Box 6248, Providence, RI 02940-6248.

Other Titles in This Series

(*Continued from the front of this publication*)

(See the AMS catalog for earlier titles)